SAFETY INSTRUMENTED SYSTEMS: Design, Analysis, and Justification

2nd Edition

By Paul Gruhn, P.E., CFSE
and
Harry Cheddie, P.Eng., CFSE

Notice

The information presented in this publication is for the general education of the reader. Because neither the author nor the publisher have any control over the use of the information by the reader, both the author and the publisher disclaim any and all liability of any kind arising out of such use. The reader is expected to exercise sound professional judgment in using any of the information presented in a particular application.

Additionally, neither the author nor the publisher have investigated or considered the affect of any patents on the ability of the reader to use any of the information in a particular application. The reader is responsible for reviewing any possible patents that may affect any particular use of the information presented.

Any references to commercial products in the work are cited as examples only. Neither the author nor the publisher endorse any referenced commercial product. Any trademarks or tradenames referenced belong to the respective owner of the mark or name. Neither the author nor the publisher make any representation regarding the availability of any referenced commercial product at any time. The manufacturer's instructions on use of any commercial product must be followed at all times, even if in conflict with the information in this publication.

Printed in the United States of America.
10 9 8 7 6 5 4 3

ISBN-13 978-1-55617-956-3
ISBN-10 1-55617-956-1

Library of Congress Cataloging-in-Publication Data

Gruhn, Paul.
Safety instrumented systems :design, analysis, and justification / by Paul Gruhn and Harry Cheddie. -- 2nd ed.
p. cm.
Includes bibliographical references.
ISBN 1-55617-956-1 (pbk.)
1. System safety. 2. Process control. 3. Industrial safety.
I. Cheddie, Harry. II. Title.
TA169.7.G78 2006
620.8'6--dc22 2005019336

TABLE OF CONTENTS

ABOUT THE AUTHORS

Harry L. Cheddie, P.Eng., CFSE

Harry Cheddie is principal engineer and partner with Exida. He is presently responsible for completing safety studies for end users, developing training programs, and teaching safety courses with an emphasis on IEC 61508 and IEC 61511.

Prior to joining Exida, Harry was a control systems advisor for Bayer Inc. in Sarnia, Ontario, Canada, where he was also the supervisor of the Central Engineering Group responsible for process control systems design and maintenance.

Harry graduated from Salford University in the U.K. with a B.Sc. (1st class honors) degree in electrical engineering. He is a registered professional engineer in the province of Ontario, Canada.

Harry is certified by the American Society for Quality as a quality engineer, and as a reliability engineer. He is also a TÜV certified functional safety expert.

Paul Gruhn, P.E., CFSE

Paul Gruhn is a safety product specialist with ICS Triplex in Houston, Texas.

Paul is an ISA Fellow. He is a member of the ISA SP84 committee, which wrote the 1996 ("Application of Safety Instrumented Systems for the Process Industries") and 2004 ("Functional Safety: Safety Instrumented Systems for the Process Industry Sector") versions of the ISA 84 series standards. Paul is the developer and instructor for ISA's three-day course EC50, "Safety Instrumented Systems," along with the matching one-day course and three-part web seminar series. He is actively involved with ISA in various local and national roles.

Paul is also a member of the System Safety Society and the National Society of Professional Engineers.

Paul has a B.S. degree in mechanical engineering from Illinois Institute of Technology in Chicago, Illinois; is a licensed professional engineer in Texas; and a certified functional safety expert (a TÜV certification).

1

INTRODUCTION

Chapter Highlights

"Engineering responsibility should not require the stimulation that comes in the wake of catastrophe."

— S. C. Florman

1.1 What Is a Safety Instrumented System?

Safety interlock system, safety instrumented system, safety shutdown system, emergency shutdown system, protective instrument system—the assorted names go on and on! Different companies within the process industry still use a variety of names for these systems. Within the ISA SP84 committee there was continual discussion (and constant changes) over the term used to describe these systems. The most generic term might be considered *safety system*, but this means different things to different people. For many chemical engineers, "safety systems" refer to management procedures and practices, not control systems. One very common term has been *emergency shutdown system* (ESD), but to electrical engineers ESD means electro-static discharge. Many don't want the word *emergency* in the name at all, as it tends to have a negative connotation. Others don't like the word 'safety shutdown system' for the same reason. Anything appearing in print with the phrase 'safety' draws immediate attention.

When the American Institute of Chemical Engineers, Center for Chemical Process Safety (AIChE CCPS) published "Guidelines for Safe Automation of Chemical Processes" in 1993, the term it used was *safety interlock system*—SIS. Some members of the ISA SP84 committee felt that *interlocks* were only one subset of many different types of safety control systems.

The ISA committee settled on the term *safety instrumented system* in order to keep the same acronym used in the AIChE text—SIS. A related AIChE CCPS text titled "Layer of Protection Analysis" released in 2001 also uses the acronym SIS, but uses the more recent definition of "safety instrumented system."

So just what *is* a safety instrumented system? The ANSI/ISA-91.00.01-2001 (Identification of Emergency Shutdown Systems and Controls That Are Critical to Maintaining Safety in Process Industries) uses the phrase *emergency shutdown system* with the following definition, "Instrumentation and controls installed for the purpose of taking the process, or specific equipment in the process, to a safe state. This does not include instrumentation and controls installed for non-emergency shutdowns or routine operations. Emergency shutdown systems may include electrical, electronic, pneumatic, mechanical, and hydraulic systems (including those systems that are programmable)." In other words, safety instrumented systems are designed to respond to conditions of a plant, which may be hazardous in themselves, or if no action were taken could eventually give rise to a hazardous event. They must generate the correct outputs to prevent or mitigate the hazardous event.

The international community has other ways of referring to these systems. International Electrotechnical Commission Standard 61508: Functional safety of electrical/electronic/programmable electronic safety-related systems (IEC 61508) uses the term safety related systems, but also introduces the combined acronym E/E/PES. As used in the title, E/E/PES stands for electric, electronic and programmable electronic. In other words, relay, solid-state, and software-based systems.

The standards generally focus on systems related to personnel safety. However, the same concepts apply to systems designed to protect equipment and the environment. After all, there are more things at risk to a company than just people. Similarly, while this text focuses on personnel safety-related systems, many of the concepts can be utilized when addressing asset and environmental applications.

As with any subject, there are a variety of acronyms and technical terms. Some terms to not have complete agreement or common usage in industry and different texts. This naturally adds to the confusion. Unless otherwise noted, all the terms used in this text are defined in ANSI/ISA-84.00.01-2004, Part 1, Clause 3. Acronyms are typically defined the first time they are used and other terms are explained where appropriate.

1.2 Who This Book Is For

This book is intended for the thousands of professionals employed in the process industries who are involved with safety systems in any way and who are expected to follow the appropriate industry standards. These individuals are employed by end users, engineering firms, system integrators, consultants, and vendors. Managers and sales individuals will also benefit from a basic understanding of the material presented.

The 1996 version of the ISA SP84's standard defined the intended audience as those who are involved in areas of "design and manufacture of SIS products, selection, and application, installation, commissioning, and pre-startup acceptance testing, operation, maintenance, documentation, and testing." Basically, if you're involved with safety systems in any way, there are portions of the standards and this book of interest to you.

The 1996 version of the standard also defined the process industry sector as, "those processes involved in, but not limited to, the production, generation, manufacture, and/or treatment of oil, gas, wood, metals, food, plastics, petrochemicals, chemicals, steam, electric power, pharmaceuticals, and waste material(s)."

The 2004 version of the ISA SP84's standard is now a global standard. It has world-wide approval and acceptance for any country utilizing IEC 61511 or ANSI/ISA-84.00.01-2004 as their national process sector functional safety standard. The ISA SP84 worked with the IEC 61511 committee to accomplish this objective. IEC 61511 and ANSI/ISA-84.00.01-2004 are identical except that ANSI/ISA-84.00.01-2004 has a grandfather clause added to it (Part 1, Clause 1). IEC 61511 and ANSI/ISA-84.00.01-2004 are clearly intended for end-users. IEC 61508 is focused for equipment manufacturers. The focus of this text is on ISA-84.00.01-2004, Parts 1-3 (IEC 61511 Mod).

1.3 Why This Book Was Written

We're engineering industrial processes—and using computer-based systems to control them—that have the potential for large-scale destruction. Single accidents are often disastrous and result in multiple fatalities and significant financial losses. We simply do not have the luxury of learning from trial and error. ("Oops, we blew up that unit and killed 20 people. Let's rebuild it, raise the set point five degrees and see what happens next time.") We must try to anticipate and prevent accidents *before* they occur. This has been one of the hard lessons learned from past accidents and why various process safety legislation was passed in different parts of the world. Hopefully this book, in its own little way, will help make the world a safer place.

The authors believe this to be the only all encompassing text on this subject. This book is a practical "how to" on the specification, analysis, design, installation and maintenance of safety instrumented systems. It includes practical knowledge needed to apply safety instrumented systems. It will hopefully serve as a guide for implementing the procedures outlined in various standards.

Aren't the standards alone enough? The answer depends upon you and your company's knowledge and experience. The "normative" (mandatory) portion of ANSI/ISA-84.01-1996 was only about 20 pages long. (There were about 80 pages of annexes and informative material.) While committee members knew what certain phrases and requirements meant, not everyone else did. Some committee members wanted certain wording specifically vague in order to have the freedom to be able to implement the requirements in different ways. Others wanted clear-cut prescriptive requirements. ANSI/ISA-84.00.01-2004 (IEC 61511 Mod) contains much more detail. Part 1 of the standard—the normative portion—is over 80 pages in length. Part 2—the 'informative' portion on how to implement Part 1—is over 70 pages. The committee felt additional material was *still* needed. At the time of this writing (early 2005), technical report ISA-TR84.00.04—Guidelines on the Implementation of ANSI/ISA-84.00.01-2004 (IEC 61511 Mod)—consists of over 200 pages of further detail. The technical report was deemed necessary as the normative and informative portions of the standard did not include the level of detail to satisfy many of the members. Such is the reality of committee work with several dozen active members and several hundred corresponding members! The two authors co-writing this text did not have the typical committee conflict issues to deal with. This is not to imply that this text is any more correct or thorough than the standards or their accompanying technical reports.

This book covers the entire lifecycle of safety instrumented systems, from determining what sort of systems are required through decommissioning. It covers the difference between process control and safety control, the separation of control and safety, independent protection layers, determining safety integrity levels, logic system and field device issues, installation, and maintenance. The book focuses on establishing design requirements, analysis techniques, technology choices, purchase, installation, documentation and testing of safety instrumented systems. It also covers the technical and economic justification for safety instrumented systems. The focus throughout is on real-world, practical solutions with many actual examples, and a minimum of theory and math. What equations are presented only involve simple algebra.

1.4 Confusion in the Industry

One goal of this book is to clarify the general confusion in the industry over the myriad choices involved in the design of safety systems. Many would have hoped to turn to industry standards for their recommendations. However, the standards are performance oriented and not prescriptive, so there are no specific recommendations. The standards essentially state what needs to be done, not specifically how to do it. For example, what follows are just a few of the choices that need to be made:

1.4.1 Technology Choices

What technology should be used; relay, solid state, or microprocessor? Does this depend on the application? Relay systems are still common for small applications, but would you want to design and wire a 500 I/O (input/output) system with relays? Is it economical to do a 20 I/O system using a redundant programmable system? Some people prefer not to use software-based systems in safety applications at all, others have no such qualms. Are some people "right" and others "wrong"?

Many feel that the use of redundant PLCs (Programmable Logic Controller) as the logic solver is the be all and end all of satisfying the system design requirements. But what about the *programming* of the PLCs? The same individuals and procedures used for programming the control systems are often used for the safety systems. Should this be allowed?

1.4.2 Redundancy Choices

How redundant, if at all, should a safety instrumented system be? Does this depend on the technology? Does it depend on the level of risk? If most relay systems were simplex (non-redundant), then why have triplicated programmable systems become so popular? When is a non-redundant system acceptable? When is a dual system required? When, if ever, is a triplicated system required? How is such a decision justified?

1.4.3 Field Devices

A safety system is much more than just a logic box. What about the field devices—sensors and final elements? Should sensors be discrete switches or analog transmitters? Should 'smart' (i.e., intelligent or processor-based) devices be used? When are redundant field devices required? What about partial stroking of valves? What about field buses? How often should field devices be tested?

1.4.4 Test Intervals

How often should systems be tested? Once per month, per quarter, per year, or per turnaround? Does this depend on technology? Do redundant systems need to be tested more often, or less often, than non-redundant systems? Does the test interval depend on the level of risk? Can systems be bypassed during testing, and if so, for how long? How can online testing be accomplished? Can testing be automated? How does a device's level of automatic diagnostics influence the manual test interval? Does the entire system need to be tested as a whole, or can parts be tested separately? How does one even *make* all these decisions?!

1.4.5 Conflicting Vendor Stories

Every vendor seems to be touting a different story line, some going so far as to imply that only *their* system should be used. Triplicated vendors take pride in showing how their systems outperform any others. Dual system vendors say their systems are just as good as triplicated systems. Is this possible? If one is good, is two better, and is three better still? Some vendors are even promoting *quad* redundant systems! However, at least one logic system vendor claims Safety Integrity Level (SIL) 3 certification for a *non-redundant* system. How can this even be possible considering the plethora of redundant logic systems? Who should one believe—and more importantly—*why*? How can one peer past all of the sales 'hype'? When overwhelmed with choices, it becomes difficult to decide at all. Perhaps it's easier just to ask a trusted colleague what he did!

1.4.6 Certification vs. Prior Use

Considering all the confusion, some vendors realized the potential benefit of obtaining certifications to various standards. Initially, this was done utilizing independent third parties. This had the desired effect of both proving their suitability and weeding out potential competition, although it was an expensive undertaking. However, industry standards in no way *mandate* the use of independently certified equipment. Users demanded the flexibility of using equipment that was *not* certified by third parties. How might a user prove the suitability of components or a system based on prior use and "certify" the equipment on their own? How much accumulated experience and documentation is required to verify that something is suitable for a particular application? How would you defend such a decision in a court of law? How about a vendor certifying themselves that they and their hardware meet the requirements of various standards? Considering how hard it is to find your own mistakes, does such a claim even have any credibility? The standards, annexes, technical reports and white papers address these issues in more detail.

1.5 Industry Guidelines, Standards, and Regulations

"Regulations are for the obedience of fools and for the guidance of wise men."

— *RAF motto*

One of the reasons industry writes its own standards, guidelines and recommended practices is to avoid government regulation. If industry is responsible for accidents, yet fails to regulate itself, the government may step in and do it for them. Governments usually get involved once risks are perceived to be 'alarming' by the general populace. The first successful regulatory legislation in the U.S. was passed by Congress over 100 years ago after public pressure and a series of marine steamboat boiler disasters killed thousands of people. Some of the following documents are performance—or goal—oriented, others are prescriptive.

1.5.1 HSE - PES

Programmable Electronic Systems In Safety Related Applications, Parts 1 & 2, U.K. Health & Safety Executive, ISBN 011-883913-6 & 011-883906-3, 1987

This document was the first of its kind and was published by the English Health & Safety Executive. Although it focused on software programmable systems, the concepts presented applied to other technologies as well. It covered qualitative and quantitative evaluation methods and many design checklists. Part 1—"An Introductory Guide"—is only 17 pages and was intended primarily for managers. Part 2—"General Technical Guidelines" —is 167 pages and was intended primarily for engineers. They were both excellent documents, although they did not appear to be well known outside the U.K. However, considering the material covered, they would appear to have been used as the foundation for many of the more recent documents.

1.5.2 AIChE - CCPS

Guidelines for Safe Automation of Chemical Processes, AIChE, 0-8169-0554-1, 1993

The American Institute of Chemical Engineers formed the Center for Chemical Process Safety (CCPS) after the accident in Bhopal, India. The CCPS has since released several dozen textbooks on various design and safety-related topics for the process industry. This particular text covers the design of Distributed Control Systems (DCS) and Safety Interlock Systems (SIS) and contains other very useful background information. The book took several years to write and was the effort of about a dozen individuals who were all from user companies (i.e., no vendors).

1.5.3 IEC 61508

Functional Safety - Safety Related Systems, IEC standard 61508, 1998

The International Electrotechnical Commission released this 'umbrella' standard which covers the use of relay, solid-state and programmable systems, including field devices. The standard applies to *all* industries: transportation, medical, nuclear, process, etc. It's a seven part document, portions of which were first released in 1998. The intention was that different industry groups would write their own industry-specific standards in line with the concepts presented in 61508. This has happened in at least the transportation, machinery and process industries. The process industry standard (IEC 61511) was released in 2003 and was focused for end users. The 61508 standard is now viewed as the standard for vendors to follow. For example, when a vendor gets a product certified for use in a particular Safety Integrity Level (SIL), the certification agency typically uses IEC 61508 as the basis for the approval.

1.5.4 ANSI/ISA-84.00.01-2004 (IEC 61511 Mod) & ANSI/ISA-84.01-1996

Functional Safety: Safety Instrumented Systems for the Process Industry Sector, ISA Standard 84.00.01-2004, Parts 1-3 (IEC 61511 Mod) and the previous *Application of Safety Instrumented Systems for the Process Industries,* ISA Standard 84.01-1996.

The ISA SP84 committee worked for more than 10 years developing this standard. The scope of this document underwent many changes through the years. It was originally intended as a U.S. standard focusing only on programmable logic boxes (and not the field devices). The scope eventually expanded to include other logic box technologies as well as field devices.

During the development of the ISA SP84's standard the IEC committee started on its 61508 general standard. The ISA SP84 committee believed its standard could be used as an industry-specific standard for the process industries under the scope of the IEC. The IEC developed its 61511 standard using ANSI/ISA-84.01-1996 as a starting point. In fact, the chairman of the ISA SP84 committee served as the chairman for the IEC 61511 standard.

ANSI/ISA-84.01-1996 stated it would be re-released in five-year intervals to account for new developments. Rather than rewrite the ISA SP84's standard from scratch, the committee decided to adopt the IEC 61511 standard with the addition of a 'grandfather clause' from the original 1996 version of the ISA SP84's standard. The new three-part standard is designated ANSI/ISA-84.00.01-2004, Parts 1-3 (IEC 61511 Mod).

1.5.5 NFPA 85

Boiler and Combustion Systems Hazard Code, National Fire Protection Association, 2004

NFPA 85 is the most recognized standard worldwide for combustion systems safety. This is a very prescriptive standard with specific design requirements. The standard covers:

- Single Burner Boiler Operation
- Multiple Burner Boilers
- Pulverized Fuel Systems
- Stoker Operation
- Atmospheric Fluidized-Bed Boiler Operation
- Heat Recovery Steam Generator Systems

The purpose of NFPA 85 is to provide safe operation and prevent uncontrolled fires, explosions and implosions. Some of the key requirements of this standard relate to the burner management system logic. The NFPA is not involved with the enforcement of this standard. However, insurance companies, regulatory agencies, and company standards often require compliance. Many countries and companies require compliance with NFPA 85 for burner management systems.

There is considerable debate as to whether a Burner Management System (BMS) is a Safety Instrumented System. There are naturally those that believe it is (as the definitions of both systems are very similar). The NFPA standard does not address Safety Integrity Levels. However, members of the various standards committees are trying to harmonize the various standards.

1.5.6 API RP 556

Recommended Practice for Instrumentation and Control Systems for Fired Heaters and Steam Generators, American Petroleum Institute, 1997

This recommended practice has sections covering shutdown systems for fired heaters, steam generators, carbon monoxide or waste gas steam generators, gas turbine exhaust fired steam generators, and unfired waste heat steam generators. While intended for use in refineries, the document states that it is "applicable without change in chemical plants, gasoline plants, and similar installations."

1.5.7 API RP 14C

Recommended Practice for Design, Installation, and Testing of Basic Surface Safety Systems for Offshore Production Platforms, American Petroleum Institute, 2001

This prescriptive recommended practice is based on "proven practices" and covers the design, installation, and testing of surface safety systems on offshore production platforms. It is intended for design engineers and operating personnel.

1.5.8 OSHA (29 CFR 1910.119 - Process Safety Management of Highly Hazardous Chemicals)

The process industry has a vested interest in writing their own industry standards, guidelines, and recommended practices. As stated earlier, if industry were to be viewed as being unable to control their own risks, there would be the possibility of government intervention. This, in fact, happened due to several significant process plant disasters in the U.S. during the 80s and 90s. 29 CFR 1910.119 was released in 1992 and, as the name implies, is directed at organizations dealing with highly hazardous substances. OSHA estimates over 25,000 facilities in the U.S. are impacted by this regulation—much more than just refineries and chemical plants. There are over a dozen sections to this legislation. A number of the sections have requirements specifically detailing issues related to the selection, design, documentation, and testing of safety instrumented systems.

For example:
Section d3: Process safety information: Information pertaining to the equipment in the process... (including) safety systems... "For existing

equipment...the employer shall *determine and document* that the equipment is designed, maintained, inspected, tested, and operating in a *safe manner*." (Emphasis added.)

People tend to have more questions *after* reading the OSHA document than before. For example, just what is 'a safe manner'? How does one 'determine', and in what way does one 'document', that things are operating 'safely'. How safe is safe enough? The OSHA document does little to answer these questions. This statement in the OSHA regulation is the basis for the 'grandfather clause' in the ISA SP84's standard. The previously mentioned standards and guidelines *do* address these issues in more detail.

Section j: Mechanical integrity: Applies to the following process equipment: ..., emergency shutdown systems, ... *Inspection and testing:* "The frequency of inspections and test of process equipment shall be consistent with applicable manufacturer's recommendations and good engineering practices, and more frequently if determined to be necessary by prior operating experience." Whose experience?! Whose good engineering practices?! The previously mentioned standards and guidelines address these issues in more detail as well.

Section j5: Equipment deficiencies: "The employer shall correct *deficiencies* in equipment that are outside *acceptable limits* before further use or in a safe and timely manner when necessary means are taken to *assure safe operation*." (Emphasis added.) What is the definition of a 'deficiency'? This sentence would seem to contradict itself. It first introduces the idea of 'acceptable limits'. (If I stand 'here', it's acceptable, but if I step over an imaginary boundary and stand over 'there', it's no longer acceptable.) This seems harmless enough. But the very same sentence then goes on to say that if anything goes wrong, you obviously didn't 'assure' (guarantee) safe operation. In other words, no matter what happens, you *can't win*. OSHA's 'general duty' clause can always be brought into play if anything goes wrong and people are injured.

Section j6: Quality assurance: "In the construction of new plants and equipment, the employer shall *assure* that equipment as it is fabricated is *suitable* for the process application for which they will be used." (emphasis added) The employer shall *'assure'*?! Benjamin Franklin said the only thing we can be 'sure' of is death and taxes. 'Suitable'?! According to whom?! The vendor trying to sell you his system? Measured against what? The industry standards address these issues in more detail.

Appendix C: Compliance guidelines and recommendations: Mechanical integrity: "Mean time to failure of various instrumentation and equipment parts would be known from the manufacturer's data or the employer's experience with the parts, which would then influence the inspection and testing frequency and associated procedures." Hopefully companies are aware that they are expected to be keeping records of this sort of information. Just how would this 'influence' the test frequency of various systems? How does one even make this determination? Some manufacturers have and do provide failure rate data, some do not. Again, the industry standards address these issues in more detail.

It's worth noting that OSHA addressed a letter to ISA in 2000 stating that it recognizes ANSI/ISA-84.01-1996 as "a recognized and generally accepted good engineering practice for SIS" and that if a company is in compliance with the standard "the employer will be considered in compliance with OSHA PSM requirements for SIS."

1.6 Standards Are Changing Their Direction

Most people want a simple "cookbook" of pre-planned solutions. For example: "For a high pressure shutdown on a catalytic cracker in a refinery, turn to page 35. There it shows dual sensor, dual logic box, non-redundant valves, yearly test interval, suggested logic programming, etc. For a high level shutdown on a high pressure separator on an unmanned offshore platform, turn to page 63. There it shows…" There are reasons the standards will never be written this way. The standards do *not* give clear, simple, precise answers. They do *not* mandate technology, level or redundancy, or test intervals.

Prescriptive standards, while helpful, cannot cover all of the variation, complexities, and details of today's systems. For example, if you purchase a pressure switch at your local hardware store, the switch will likely satisfy the requirements of certain prescriptive standards. However, there will be little, if any, requirements about how *well* the components have to perform.

Similarly, twenty years ago most safety systems consisted of discrete switches, discrete relay logic, and on-off valves controlled by solenoids. Things were much simpler back then. Sensors today may be discrete switches, conventional analog transmitters, smart transmitters, or safety transmitters. Logic solvers may now be relay logic, solid-state logic, conventional PLCs, or safety PLCs. Final elements may now be on/off valves with solenoids or control valves with smart positioners. Prescriptive standards simply cannot address the selection of such a diverse array of components and technology. However, newer performance based standards *do* provide the means to make the correct selections.

There is a fundamental change in the way industry standards are being written. Standards are moving away from *prescriptive* standards and toward more *performance*-oriented requirements. In fact, this was one of the recommendations made in a government report after the Piper Alpha offshore platform explosion in the North Sea. Prescriptive standards generally do not account for new developments or technology and can easily become dated. This means each organization will have to decide for themselves just what is 'safe'. Each organization will have to decide how they will 'determine' and 'document' that their systems are, in fact, 'safe'. Unfortunately, these are difficult decisions that few want to make, and fewer still want to put in writing. "What is safe" transcends pure science and deals with philosophical, moral, and legal issues.

1.7 Things Are Not As Obvious As They May Seem

Intuition and gut feel do not always lead to correct conclusions. For example, which system is safer, a dual one-out-of-two system (where only one of the two redundant channels is required in order to generate a shutdown) or a triplicated two-out-of-three system (where two of the three redundant channels are required in order to generate a shutdown)? Intuition might lead you to believe that if one system is "good," two must be better, and three must be the best. You might therefore conclude that the triplicated system is safest. Unfortunately, it's not. It's very easy to show that the dual system is actually safer. Chapter 8 will deal with this subject in more detail. However, for every advantage there is a disadvantage. The one-out-of-two system may be safer, but will suffer more nuisance trips. Not only does this result in lost production downtime and economic issues, it is generally recognized that there is nothing "safe" about nuisance trips, even though they are called "safe failures."

At least two recent studies, one by a worldwide oil company, another by a major association, found that a significant portion of existing safety instrumented functions were both over-designed (37-49%), as well as under-engineered (4-6%). Apparently things are not as obvious as people may have thought in the past. The use of performance-based standards should allow industry to better identify risks and implement more appropriate and cost effective solutions.

If there hasn't been an accident in your plant for the last 15 years, does that mean that you have a safe plant? It might be tempting to think so, but nothing could be further from the truth. You may not have had a car accident in 15 years, but if you've been driving home every night from a bar after consuming 6 drinks, I'm not about to consider you a "safe" driver! No doubt people may have made such statements one day before Seveso

(Italy), Flixborough (England), Bhopal (India), Chernobyl (Soviet Union), Pasadena (USA), etc. Just because it hasn't happened yet, doesn't mean it won't, or can't.

If design decisions regarding safety instrumented systems were simple, obvious, and intuitive, there would be no need for industry standards, guidelines, recommended practices, or this book. Airplanes and nuclear power plants are *not* designed by intuition or gut feel. How secure and safe would you feel if you asked the chief engineer of the Boeing 777, "Why did you choose that size engine, and only two at that?", and his response was, "That's a good question. We really weren't sure, but that's what our vendor recommended." You'd like to think that Boeing would know how to engineer the entire system. Indeed they do! Why should safety instrumented systems be any different? Do you design all of your systems based on your vendor's recommendations? How would you handle conflicting suggestions? Do you really want the fox counting your chickens or building your henhouse?

Many of the terms used to describe system performance seem simple and intuitive, yet they've been the cause of much of the confusion. For example, can a system that's 10 times more "reliable" be less "safe"? If we were to replace a relay-based shutdown system with a newer PLC that the vendor said was 10 times more "reliable" than the relay system, would it automatically follow that the system was safer as well? Safety and reliability are *not* the same thing. It's actually very easy to show that one system may be more "reliable" than another, yet still be *less safe*.

1.8 The Danger of Complacency

It's easy to become overconfident and complacent about safety. It's easy to believe that we as engineers using modern technology can overcome almost any problem. History has proven, however, that we cause our own problems and we always have more to learn. Bridges will occasionally fall, planes will occasionally crash, and petrochemical plants will occasionally explode. That does *not* mean, however, that technology is bad or that we should live in the Stone Age. It's true that cavemen didn't have to worry about The Bomb, but then we don't have to worry about the plague. We simply need to learn from our mistakes and move on.

After Three Mile Island (the worst U.S. nuclear incident), but before Chernobyl (the worst nuclear incident ever), the head of the Soviet Academy of Sciences said, "Soviet reactors will soon be so safe that they could be installed in Red Square." Do you think he'd say that *now*?

The plant manager at Bhopal, India was not in the plant when that accident happened. When he was finally located, he could not accept that his plant was actually responsible. He was quoted as saying "The gas leak just can't be from my plant. The plant is shut down. Our technology just can't go wrong. We just can't have leaks." One wonders what he does for a living now.

After the tanker accident in Valdez, Alaska, the head of the Coast Guard was quoted as saying, "But that's impossible! We have the perfect navigation system?"

Systems can always fail; it's just a matter of when. People can usually override any system. Procedures will, on occasion, be violated. It's easy to become complacent because we've been brought up to believe that technology is good and will solve our problems. We want to have faith that those making decisions know what they're doing and are qualified. We want to believe that our 'team' is a 'leader', if for no other reason than the fact that we're on it.

Technology may be a good thing, but it is not infallible. We as engineers and designers must never be complacent about safety.

1.9 There's Always More to Learn

There are some who are content to continue doing things the way they've always done. "That's the way we've done it here for 15 years and we haven't had any problems! If it ain't broke, don't fix it."

Thirty years ago, did we know all there was to know about computers and software? If you brought your computer to a repair shop with a problem and found that their solution was to reformat the hard drive and install DOS as an operating system (which is what the technician learned 15 years ago), how happy would you be?

Thirty years ago, did we know all there was to know about medicine? Imagine being on your death bed and being visited by a 65-year-old doctor. How comfortable would you feel if you found out that that particular doctor hadn't had a single day of continuing education since graduating from medical school 40 years ago?

Thirty years ago, did we know all there was to know about aircraft design? The Boeing 747 was the technical marvel 30 years ago. The largest engine we could make back then was 45,000 pounds thrust. We've learned a lot since then about metallurgy and engine design. The latest generation engines can now develop over 100,000 pounds thrust. It no longer takes four engines to fly a jumbo jet. In fact, the Boeing 777, which has replaced many 747s at some airlines, only has two engines.

Would you rather learn from the mistakes of others, or make them all yourself? There's a wealth of knowledge and information packed into recent safety system standards as well as this textbook. Most of it was learned the hard way. Hopefully others will utilize this information and help make the world a safer place.

So now that we've raised some of the issues and questions, let's see how to answer them.

Summary

Safety instrumented systems are designed to respond to the conditions of a plant, which may be hazardous in themselves, or if no action is taken could eventually give rise to a hazardous event. They must generate the correct outputs to prevent or mitigate the hazardous event. The proper design and operation of such systems are described in various standards, guidelines, recommended practices, and regulations. The requirements, however, are anything but intuitively obvious. Setting specifications, selecting technologies, levels of redundancy, test intervals, etc. is not always an easy, straightforward matter. The various industry standards, as well as this book, are written to assist those in the process industries tasked with the proper selection, design, operation, and maintenance of these systems.

References

1. *Programmable Electronic Systems in Safety Related Applications - Part 1 - An Introductory Guide*. U.K. Health & Safety Executive, 1987.
2. *Guidelines for Safe Automation of Chemical Processes*. American Institute of Chemical Engineers - Center for Chemical Process Safety, 1993.
3. ANSI/ISA-84.00.01-2004, Parts 1-3 (IEC 61511-1 to 3 Mod). *Functional Safety: Safety Instrumented Systems for the Process Industry Sector* and ISA-84.01-1996. *Application of Safety Instrumented Systems for the Process Industries.*
4. IEC 61508-1998. *Functional Safety of Electrical/Electronic/Programmable Electronic Safety-Related Systems.*

5. 29 CFR Part 1910.119. *Process Safety Management of Highly Hazardous Chemicals*. U.S. Federal Register, Feb. 24, 1992.

6. Leveson, Nancy G. *Safeware - System Safety and Computers.* Addison-Wesley, 1995.

2
DESIGN LIFECYCLE

Chapter Highlights

"If I had 8 hours to cut down a tree, I'd spend 6 hours sharpening the axe."

— A. Lincoln

Designing a single component may be a relatively simple matter, one that a single person can handle. Designing any large *system* however, whether it's a car, a computer, or an airplane, is typically beyond the ability of any single individual. The instrument or control system engineer should *not* feel that all the tasks associated with designing a safety instrumented system are his or her responsibility alone, because they're not. The design of a system, including a safety instrumented system, requires a multi-discipline *team*.

2.1 Hindsight/Foresight

"Hindsight can be valuable when it leads to new foresight."

— P. G. Neumann

Hindsight is easy. Everyone always has 20/20 hindsight. *Fore*sight, however, is a bit more difficult. Foresight is required, however, with today's large, high risk systems. We simply can't afford to design large petrochemical plants by trial and error. The risks are too great to learn that way. We have to try and prevent certain accidents, no matter how remote the possibility, even if they have never yet happened. This is the subject of *system safety*.

System safety was born out of the military and aerospace industries. The military have many obvious high risk examples. The following case may have been written in a lighthearted fashion, but was obviously a very serious matter to the personnel involved. Luckily, there were no injuries.

> *An ICBM silo was destroyed because the counterweights, used to balance the silo elevator on the way up and down, were designed with consideration only to raising a fueled missile to the surface for firing. There was no consideration that, when you were not firing in anger, you had to bring the fueled missile back down to defuel. The first operation with a fueled missile was nearly successful. The drive mechanism held it for all but the last five feet when gravity took over and the missile dropped back down. Very suddenly, the 40-foot diameter silo was altered to about 100-foot diameter. [Ref. 1]*

A radar warning system in Greenland suffered an operational failure in the first month. It reported inbound Russian missiles, but what it actually responded to was...*the rising moon*.

If you make something available to someone, it will at some point be used, even if you didn't intend it to be. For example, there were two cases where North American Radar Defense (NORAD) and Strategic Air Command (SAC) went on alert because radar systems reported incoming missiles. In reality, someone had just loaded a training tape by mistake. After the same incident happened a *second* time, it was finally agreed upon to store the training tapes in a different location! What might have originally been considered human error was actually an error in the *system* which allowed the inevitable human error to happen.

2.2 Findings of the HSE

The English Health and Safety Executive examined 34 accidents that were the direct result of control and safety system failures in a variety of different industries [Ref. 2]. Their findings are summarized in Figure 2-1. Most accidents could have been prevented. The majority of accidents (44%) were due to *incorrect and incomplete specifications*. Specifications consist of both the *functional* specification (i.e., *what* the system should do) and the *integrity* specification (i.e., how *well* it should do it).

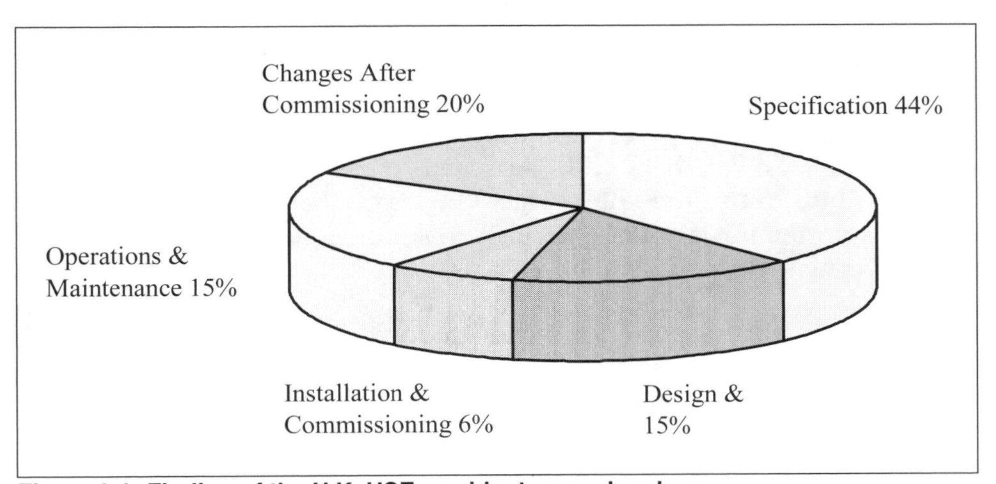

Figure 2-1: Finding of the U.K. HSE, accident cause by phase

There are many examples of functional specification errors. Kletz [Ref. 3] documented a case where a computer controlled an exothermic reactor. When material was added to the reactor the flow of cooling water needed to increase. However, the system was also programmed so that for any fault in the plant—and many things were categorized as a fault—the outputs would freeze at their last known value. Fate would have it that these two conflicting conditions happened at the same time. Material was added to the reactor and then the system detected a low gear box oil level. The flow of cooling water did not increase so the reactor overheated and discharged its contents. The system did exactly what it was programmed to do. There was no hardware failure.

One Mars probe was designed to circle the planet, yet it crashed into the planet instead. It was determined that the U.S. consortium worked with English units, but the European consortium worked with metric units. There was no hardware failure.

Another Mars probe was designed to land on the planet. Landing legs were designed to deploy at a certain altitude and the rocket motor would then shut off. The mechanical designers knew that the deployment of the landing legs created spurious signals that needed to be temporarily ignored. Unfortunately, this requirement was not documented or passed on to others. The software designers simply programmed the computer to shut off the rocket motor once operation of the landing legs had been detected. The result was that the rocket shut off at too high of an altitude and the satellite was destroyed in a crash landing. There was no hardware failure. A simple omission in the specification resulted in the loss of the satellite which cost over $100 million, not even considering the cost of the booster rocket.

You make shake your head over the silliness of these cases, but there are literally *thousands* of them documented in various texts. There is much to learn from them, even if they happened in an industry different than yours. Trevor Kletz said, "Accidents are not due to lack of knowledge, but failure to use the knowledge we already have." Would you rather learn from the mistakes of others, or make them all yourself, especially if your very life depended on it?

The integrity specification has to do with system performance. The HSE documented a case where a user replaced a relay system with a Programmable Logic Controller (PLC). They implemented the same functional logic and therefore assumed it was the same system—even improved through the use of a newer technology. Unfortunately, the newer technology had completely different failure characteristics than the older system and was actually *less* safe. More will be covered about this topic in subsequent chapters.

Leveson [Ref. 4] has reported that in the vast majority of accidents in which software was involved, flaws could be traced back to the requirements—in other words, incomplete or wrong assumptions about the operation of the controlled system or required operation of the computer, unhandled system states, environmental conditions, etc.

Some accidents have happened because systems were used in ways unimagined by the original system designers. For example, one Patriot missile missed an incoming Scud during the first Gulf war. The impact of the Scud at a U.S. facility resulted in multiple fatalities. The Patriot missile system was originally intended and designed for relatively short duration operation. It was not intended to be left on for long periods of time. When operated in such a manner, internal counters and timers differed enough to impact the missile's accuracy. This was the result of the system being continually modified over the years and using different numbering systems in over 20 programming languages. There was no hardware failure.

Needless to say, only the system *user* can develop system specifications. No vendor can possibly imagine or account for all modes of operation. No vendor can tell a user how best to operate their facility or what sort of system logic would be most appropriate.

The next largest portion of problems (20%) found by the HSE were due to changes made after commissioning. What one person thought was a minor, insignificant issue was not properly documented or reviewed by appropriate personnel. Missiles have blown off course and process plants have exploded due to such problems. The end user is primarily responsible for changes made after commissioning.

Operations and maintenance problems were found to be responsible for 15% of accidents. There have been cases where technicians fixed one item, only to introduce a problem elsewhere in the system. In other cases they thought they fixed the item under consideration, yet they actually didn't. End users are primarily responsible for operation and maintenance.

Design and implementation errors accounted for 15% of problems. This is about the only error that falls within the realm of the vendor or system integrator. There have been cases where specifications were correct, but the system supplied did not meet at least one of the requirements and was not thoroughly tested in order to reveal the flaw.

A relatively small percentage of errors (6%) were traced back to installation and commissioning issues.

Overall, similar finding have been published by other organizations, although not necessarily down to the level of detail of actual percentages for each failure classification.

In order to design a safe system, one needs to consider all of these different areas and not just focus on the one that may be the easiest or most convenient to cover. The committees writing the different industry standards and guidelines realized this and attempted to cover all of the bases. This text will attempt to do the same.

2.3 Design Lifecycle

Large systems require a methodical design *process* to prevent important items from falling through cracks. Figure 2-2 shows the lifecycle steps as described in the ANSI/ISA-84.00.01-2004 (IEC 61511 Mod) standard. This should be considered one example only. There are variations of the lifecycle presented in other industry documents. A simplified lifecycle is shown in Figure 2-3. A company may wish to develop their own variation of the lifecycle based on their unique requirements.

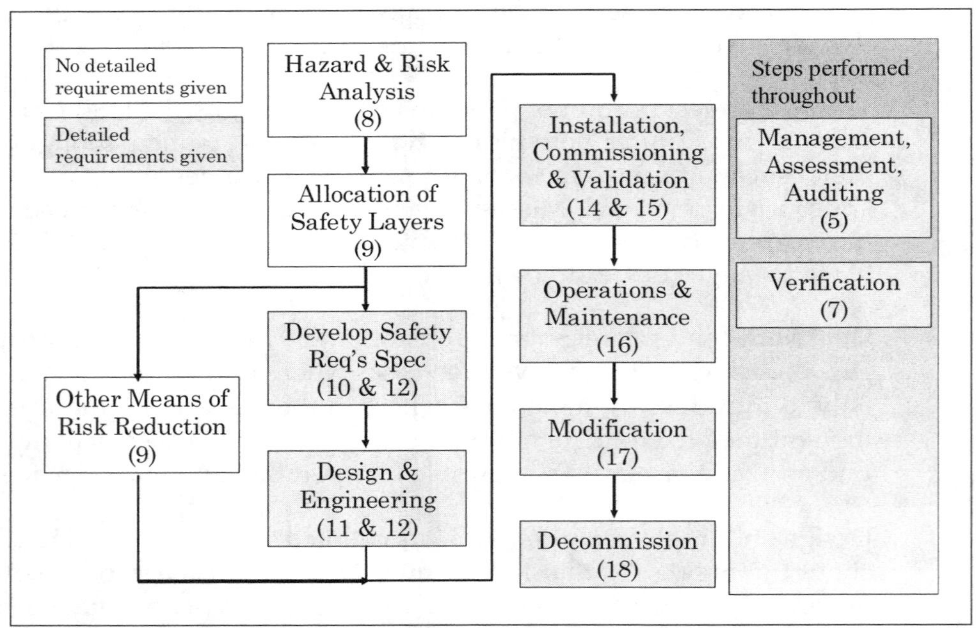

Figure 2-2: SIS Design Lifecycle (with Clause numbers from IEC 61511)

Some will complain that performing all of the steps in the life cycle, like all other tasks designed to lower risk, will increase overall costs and result in lower productivity. One in-depth study in the past [Ref. 4], conducted by a group including major engineering societies, 20 industries, and 60 product groups with a combined exposure of over 50 billion hours, concluded that *production increased as safety increased*. In the U.S., OSHA (Occupational Safety and Health Administration) documented that since the adoption of their process safety management regulation, the number of accidents has decreased over 20% and companies are reporting that their productivity is *higher*.

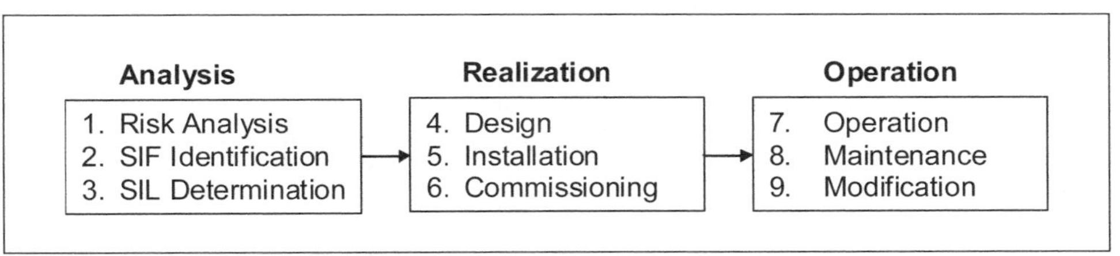

Figure 2-3: Simplified Lifecycle

2.3.1 Hazard & Risk Analysis

One of the goals of process plant design is to have a facility that is inherently safe. As Trevor Kletz said, "What you don't have, can't leak." Hopefully the design of the process can eliminate many of the hazards, such as eliminating unnecessary storage of intermediate products, using safer catalysts, etc. Entire books have been devoted to the subject of inherently safe process designs.

The fist step described in the standard is developing an understanding of the hazards and risks associated with the process. *Hazard analysis* consists of *identifying* the hazards and hazardous events. There are numerous techniques that can be used (e.g., HAZOP, What If, Fault Tree, Checklist, etc.) and many texts to describe each method. *Risk assessment* consists of *ranking* the risk of the hazardous events that have been identified in the hazard analysis. Risk is a function of the frequency or probability of an event, and the severity or consequences of the event. Risks may impact personnel, production, capital equipment, the environment, company image, etc. Risk assessment can either be qualitative or quantitative. Qualitative assessments subjectively rank the risks from low to high. Quantitative assessments, as the name obviously implies, attempt to assign numerical factors to the risk, such as death or accident rates, actual size of a release, etc. These studies are *not* the sole responsibility of the instrument or control system engineer. There are obviously a number of other disciplines required in order to perform these assessments, such as safety, operations, maintenance, process, mechanical design, electrical, etc.

2.3.2 Allocation of Safety Functions to Protective Layers

If the risks associated with a hazardous event can be prevented or mitigated with something other than instrumentation—which is complex, expensive, requires maintenance, and is prone to failure—so much the better. For example, a dike is a simple and reliable device that can easily contain a liquid spill. KISS (Keep it Simple, Stupid) should be an overriding theme.

For all safety functions assigned to instrumentation (i.e., safety instrumented functions), the level of performance required needs to be determined. The standards refer to this as safety integrity level (SIL). This continues to be a difficult step for many organizations. Note that SIL is not directly a measure of process risk, but rather a measure of the safety system performance required in order to control the risks identified earlier to an acceptable level. The standards describe a variety of methods on how this can be done.

2.3.3 Develop Safety Requirements Specification

The next step consists of developing the safety requirements specification. This consists of documenting the input and output (I/O) requirements, functional logic, and the SIL of each safety function. This will naturally vary for each system. There is no general, across the board recommendation that can be made. One simple example might be, "If temperature sensor TT2301 exceeds 410 degrees, then close valves XV5301 and XV5302. This function must respond within 3 seconds and needs to meet SIL 2." It may also be beneficial to list reliability requirements if nuisance trips are a concern. For example, many different systems may be designed to meet SIL 2 requirements, but each will have different nuisance trip performance. Considering the costs associated with lost production downtime, as well as safety concerns, this may be an important issue. Also, one should include *all* operating conditions of the process, from start-up through shutdown, as well as maintenance. One may find that certain logic conditions conflict during different operating modes of the process.

The system will be programmed and tested according to the logic determined during this step. If an error is made here, it will carry through for the rest of the design. It won't matter how redundant or how often the system is manually tested, it simply won't work properly when required. These are referred to as systematic or functional failures. Using diverse redundant systems programmed by different people using different languages and tested by an independent team will *not* help in this situation because the functional logic they all base their work on could have an error.

A mechanical analogy may be helpful. Suppose we have a requirement for a valve. The valve must be a certain size, must close within a certain period of time, must operate with a certain pressure drop, etc. This represents the functional specification, *what* the valve must do. Different vendors can naturally implement the requirements in different ways. If a carbon steel valve is installed, and two months later the valve is found to have failed due to heavy corrosion, where was the error made? The functional specification just described *what* the valve was to do. In this case the *integrity* specification was incomplete and did not state the corrosive service and that a special material was therefore required. The integrity specification states how *well* the system should perform its function. In terms of an SIS, this would be the SIL.

2.3.4 SIS Design & Engineering

This step covers conceptual design, detail design, and factory testing. Any proposed *conceptual design* (i.e., a proposed implementation) needs to be analyzed to see if it meets the functional and performance requirements. One doesn't pick a certain size jet engine for an aircraft based on intuition. One doesn't size a million dollar compressor by "gut feel". One doesn't determine the size of pilings required for a bridge by trial and error, at least not any more. One needs to initially select a technology, configuration (i.e., architecture), test interval, etc. This pertains to the field devices as well as the logic box. Factors to consider are overall size, budget, complexity, speed of response, communication requirements, interface requirements, method of implementing bypasses, testing, etc. One can then perform a simple quantitative analysis to see if the proposed system meets the performance requirements. The intent is to evaluate the system *before* one specifies the solution. Just as it's better to perform a HAZOP *before* you build the plant rather than afterward, it's better to analyze the proposed safety system *before* you specify, build and install it. The reason for both is simple. It's cheaper, faster and easier to redesign on paper.

Detail design involves the actual documentation and fabrication of the system. Once a design has been chosen the system must be engineered and built following strict and conservative procedures. This is the only realistic method of preventing design and implementation errors that we know of. The process requires thorough documentation which serves as an auditable trail that someone else may follow for independent verification purposes. It's difficult to catch one's own mistakes.

After the system is constructed, the hardware and software should be fully tested at the integrator's facility. Any changes that may be required will be easier to implement at the factory rather than the installation site.

2.3.5 Installation, Commissioning, and Validation

This step is to ensure that the system is installed and started up according to the design requirements and performs per the safety requirements specification. The entire system must be checked, this time including the field devices. There should be detailed installation, commissioning, and testing documents outlining each procedure to be carried out. Completed checks should be signed off in writing, documenting that each and every function has been checked and has satisfactorily passed all tests.

2.3.6 Operations and Maintenance

Not all faults are self-revealing. Therefore, *every* safety instrumented system *must* be periodically tested and maintained. This is necessary in order to make certain that the system will respond properly to an actual demand. The frequency of inspection and testing will have been determined earlier in the lifecycle. All testing must be documented. This will enable an audit to determine if the initial assumptions made during the design (e.g., failure rates, failure modes, test intervals, etc.) are valid based on actual operation.

2.3.7 Modifications

As process conditions change, it may be necessary to make modifications to the safety system. All proposed changes require returning to the appropriate phase of the life cycle in order to review the impact of the change. A change that may be considered minor by one individual may actually have a major impact to the overall process. This can only be determined if the change is documented and thoroughly reviewed by a qualified team. Hindsight has shown that many accidents have been caused by this lack of review. Changes that are made must be thoroughly tested.

2.3.8 Decommissioning

Decommissioning a system entails a review to make sure removing the system from service will not impact the process or any other surrounding units. Means must be available during the decommissioning process to protect the personnel, equipment, and the environment.

Summary

The overall design of safety instrumented systems is not a simple, straightforward matter. The total engineering knowledge and skills required are often beyond that of any single person. An understanding of the process, operations, instrumentation, control systems, and hazard analysis are required. This typically calls for the interaction of a multi-disciplined team.

Experience has shown that a detailed, systematic, methodical, well documented design *process* is necessary in the design of safety instrumented systems. This starts with a safety review of the process, implementation of other safety layers, systematic analysis, as well as detailed documentation and procedures. These steps are described in various regulations, standards, guidelines, and recommended procedures. The steps are referred to as the safety design life cycle. The intent is to leave a documented, auditable trail, and to make sure that nothing is neglected or falls between the inevitable cracks within every organization.

References

1. *Air Force Space Division Handbook.*
2. *Out of Control: Why control systems go wrong and how to prevent failure.* U.K. Health & Safety Executive, 1995.
3. Kletz, Trevor A. *Computer Control and Human Error.* Gulf Publishing Co., 1995.
4. Leveson, Nancy G. *Safeware - System Safety and Computers.* Addison-Wesley, 1995.
5. ANSI/ISA-84.00.01-2004, Parts 1-3 (IEC 61511-1 to 3 Mod). *Functional Safety: Safety Instrumented Systems for the Process Industry Sector.*
6. IEC 61508-1998. *Functional Safety of Electrical/Electronic/Programmable Electronic Safety-Related Systems.*
7. *Programmable Electronic Systems in Safety Related Applications, Part 1 - An Introductory Guide.,* U.K. Health & Safety Executive, 1987.
8. *Guidelines for Safe Automation of Chemical Processes.* American Institute of Chemical Engineers - Center for Chemical Process Safety, 1993.
9. Neumann, Peter. G. *Computer Related Risks.* Addison-Wesley, 1995.

3

PROCESS CONTROL VS. SAFETY CONTROL

Chapter Highlights

"Nothing can go wrong, click... go wrong, click... go wrong, click..."

— *Anonymous*

Process control used to be performed in pneumatic, analog, single loop controllers. Safety functions were performed in different hardware, typically hardwired relay systems. Electronic distributed control systems (DCSs) started to replace single loop controllers in the 1970s. Programmable logic controllers (PLCs) were developed to replace relays in the late 1960s. Since both systems are software programmable, some people naturally concluded that there would be benefits in performing both control and safety functions within the same system, usually the DCS. The typical benefits touted included single source of supply, integrated communications, reduced training and spares, simpler maintenance, and potentially

lower overall costs. Some believe that the reliability, as well as the redundancy, of modern DCSs are "good enough" to allow such combined operation. However, all domestic and international standards, guidelines, and recommended practices clearly recommend *separation* of the two systems. The authors agree with this recommendation and wish to stress that *the reliability of the DCS is not the issue.*

3.1 Control and Safety Defined

Critical systems require testing and thorough documentation. It's debatable whether normal process control systems require the same rigor of testing and documentation. When the US government came out with their process safety management (PSM) regulation (29 CFR 1910.119) in 1992, many questioned whether the mandated requirements for documentation and testing applied to both the control systems as well as the safety systems. For example, most organizations have documented testing procedures for their safety instrumented systems, but the same may not be said for all of their control system loops. Users in the process industry questioned OSHA representatives as to whether the requirements outlined in the PSM regulation applied to all 6,000 loops in their DCS, or just the 300 in their safety instrumented systems. OSHA's response was that it included everything. Users felt this was another nail in the proverbial coffin trying to put them out of business.

This helped fuel the development of ANSI/ISA-91.01-1995 "Identification of Emergency Shutdown Systems and Controls That Are Critical to Maintaining Safety in Process Industries [Ref. 10]". The ANSI/ISA-91.00.01-2001 standard was reaffirmed in 2001. This brief standard (only two pages long) includes definitions of process control, safety control, and safety critical control.

The ANSI/ISA-91.00.01-2001 standard defines a "Basic Process Control System" (BPCS), as opposed to an 'advanced' process control system, as "the control equipment installed to perform the normal regulatory functions for the process (e.g., PID control and sequential control)." Some have stated that this accounts for up to 95% of instrumentation for most land-based facilities. Most people accomplish this with a DCS, PLC, or hybrid system.

The ANSI/ISA-91.00.01-2001 standard defines "Emergency Shutdown System" as "instrumentation and controls installed for the purpose of taking the process, or specific equipment in the process, to a safe state. This does not include instrumentation and controls installed for non-emergency shutdowns or routine operations." Some have stated that this accounts for less than 10% of instrumentation for most land-based facilities.

The ANSI/ISA-91.00.01-2001 standard also defines a third category, "Safety Critical Control," as "a control whose failure to operate properly will directly result in a catastrophic release of toxic, reactive, flammable or explosive chemical." This essentially means any control or safety layer, without additional safety layers, where a failure will immediately result in a hazardous event. Hopefully, the percentage of such loops is very small.

Based on these definitions, users stated to OSHA that their 6,000 DCS loops were not safety-related and therefore did not require the same degree of rigor for documentation and testing as their 300 safety instrumented loops.

This is not meant to imply that the design of distributed control systems does not require thorough analysis, documentation, and management controls. They obviously do, just not to the same extent as safety systems.

3.2 Process Control - Active/Dynamic

It's important to realize and understand the fundamental differences between process control and safety control. Process control systems are active, or dynamic. They have analog inputs, analog outputs, perform math and number crunching, and have feedback loops. Therefore, most failures in control systems are inherently self-revealing. For example, consider the case of a robot on an automated production line. Normally the robot picks up part A and places it in area B. If the system fails, it's obvious to everyone; it no longer places part A in area B. There's no such thing as a "hidden" failure. The system either works or it doesn't. There's only one failure mode with such systems—revealed—and you don't need extensive diagnostics to annunciate such failures. If a modulating process control valve were to fail full open or full closed, it will most likely impact production and the problem would become evident to everyone very quickly. If the valve were to freeze in its last state the resulting problems would also make themselves known relatively quickly. Again, extensive testing and diagnostics are usually not required to reveal such failures or problems.

3.2.1 The Need for (and Ease of) Making Frequent Changes

Process control systems must be flexible enough to allow frequent changes. Process parameters (e.g., set points, PID settings, manual/automatic, etc.) require changing during normal operation. Portions of the system may also be placed in bypass and the process may be controlled manually. This is all normal for control systems. Simple, easy, and quick access is a requirement in order to make such changes.

3.3 Safety Control - Passive/Dormant

Safety systems, however, are just the opposite. They're dormant, or passive. They operate for extended periods of time doing virtually nothing and hopefully will never be called into action. An example would be a pressure relief valve. Normally, the valve is closed. It only opens when the pressure reaches a certain limit. If the pressure never exceeds that value, the valve never operates. Many failures in these systems may *not* be self-revealing. If the relief valve is plugged there is no immediate indication. A PLC could be hung up in an endless loop. Without a watchdog timer, the system would not be able to recognize the problem, and if the system was dormant, neither would you. An output module might use triacs which could fail energized. Many systems are unable to recognize such a problem. If the output was energized during normal operation, how would you recognize such a condition? How confident are you that a valve that hasn't been stroked in seven years will actually close? How confident are you that a backup generator that hasn't even been turned on in two years will actually run when you need it? For that matter, how confident are you that your lawn mower will start if it's been left sitting in the garage all winter long with gasoline in the tank and carburetor? How much confidence can you place in anything that under normal circumstances is not functioning (or changing state) to begin with? The only way to know if dormant systems will work when you need them is to test them. Either you must test the system or it must be able to test itself. Hence, there is a need for extensive diagnostics in dormant, passive safety-related systems. The alternative is to use inherently fail-safe systems, where such dangerous failures are highly unlikely by design, but this is generally not possible for all components.

What this means is that systems designed for *control* may be totally unsuitable for *safety*. The two primary issues have to do with security (i.e., access control) and diagnostics.

Safety systems require a high level of access control in order to implement security features and enforce management of change procedures. Safety functions should not be tampered with. How do you provide operators with easy access to some functions, but lock them out of others? How do you know that making a change in one area of the system won't impact a critical function elsewhere? Many control systems do not implement very effective access control measures.

Extensive diagnostics are required in safety systems simply because not all failures are self-revealing. Many people *assume* that all modern electronic systems include extensive diagnostics. Unfortunately, this is *not* the case. There are very simple economic reasons why this is so. This will be covered in more detail in Chapter 7.

3.3.1 The Need for Restricting Changes

Safety instrumented systems should be designed to allow very little human interaction. Operators interact with and control the process using the control system. If control systems can no longer maintain control, independent alarms generally indicate a problem. Operators can then supervise manual operations. If automatic process control systems and human intervention cannot make the required corrections, the last line of defense should function automatically and independently. About the only interaction that should be allowed between operators and safety systems are overrides for allowing startup and maintenance work on portions of the system. Note that strict procedures must be adhered to at such times. People must know that there is a last line of defense—something that can be relied upon. These systems should have tightly controlled access. The last thing anyone wants is a system that doesn't function because some lone operator disabled the system without anyone else's knowledge. It has happened.

3.3.2 Demand Mode vs. Continuous Mode

The 1993 AIChE CCPS guideline text [Ref. 3] and the 1996 version of ANSI/ISA-84.01 [Ref. 5] considered safety systems to be low demand mode systems, although this term wasn't used at that time. The thought was that properly designed safety systems would have demands placed on them infrequently (e.g., once per year). The more recent IEC 61508 standard [Ref. 4], which was written to address *all* industries, introduced the concept of high demand—or continuous mode—systems. These are simply systems that have frequent demands placed on them. A classic example is the brakes in a car. The concept of continuous mode systems filtered down into the IEC 61511 standard for the process industry and the ISA SP84 committee.

As one might expect from any change, problems ensued. The safety system performance requirements originally documented for the different safety integrity levels (SILs), only apply for low demand mode systems. The concept of a "probability of failure on demand" does not make sense for systems that might have 10 or 10,000 demands placed on them in a year. While there are naturally some that disagree with the solution chosen, the committee adopted the concept of assigning dangerous failure *rates,* rather than probabilities, to the different SILs. The difference between the rates and probabilities is a factor of 10,000, based on the simplification that there are approximately 10,000 hours in a year.

Continuous mode systems are essentially similar in concept to what the ANSI/ISA-91.01 standard [Ref. 12] refers to as "safety critical control." Even members of the ISA SP84 committee have difficulty citing examples of continuous mode systems in the process industry. Some examples, such as a high level shutoff in a batch reactor that is loaded once per day, are considered by some people to be merely examples of poor process control, not safety.

3.4 Separation of Control and Safety Systems

An ongoing topic of controversy in the process industry is whether control and safety systems should be combined within one system. In other words, should all safety functions be combined in the process control system? Proponents would argue that both systems nowadays are programmable and that process control systems are reliable and can be redundant, so why not? The rebuttals are rather simple and do not hinge on reliability. As we shall see, all of the standards, recommended practices, and guidelines in industry recommend that *separate* systems be provided for process and safety control.

Proponents of combining control and safety also cite that since control systems are active and dynamic, and faults are inherently self-revealing (as stated earlier), that they're therefore ideally suited for use in safety since there would be no hidden faults. For example, an analog control valve is continually being stroked, which would identify problems. True, but not *all* problems. Having a separate solenoid on a control valve (powered by a separate safety system) may still not meet the overall safety requirements. Stroking a valve does not ensure that it will close or seal completely, and stroking the valve does not test that the solenoid will work when required. If the valve does fail stuck (everything fails, it's just a matter of when), then no safety action will be possible. Careful analysis is required to see if the safety requirements will be fulfilled when combining control and safety, as there are many other issues to consider.

Trevor Kletz, one of the industry leaders in terms of process safety has stated it simply and very well. "Safety systems such as emergency trips should be completely independent of the control system and, when practicable, hard-wired. If they are based on a computer, it should be independent of the control computer." [Ref. 1]

The following subsections describe what is stated in other documents.

3.4.1 HSE - PES

The English couldn't make the message of separation any more clear. Their document [Ref. 2] comes as a two part volume. Part I is for the managers and is only 17 pages long. Figure 3-1 takes an *entire page*. (The technical document—the one for the engineers—Part II, is about 170 pages.) Separate sensors, separate logic box, separate valves.

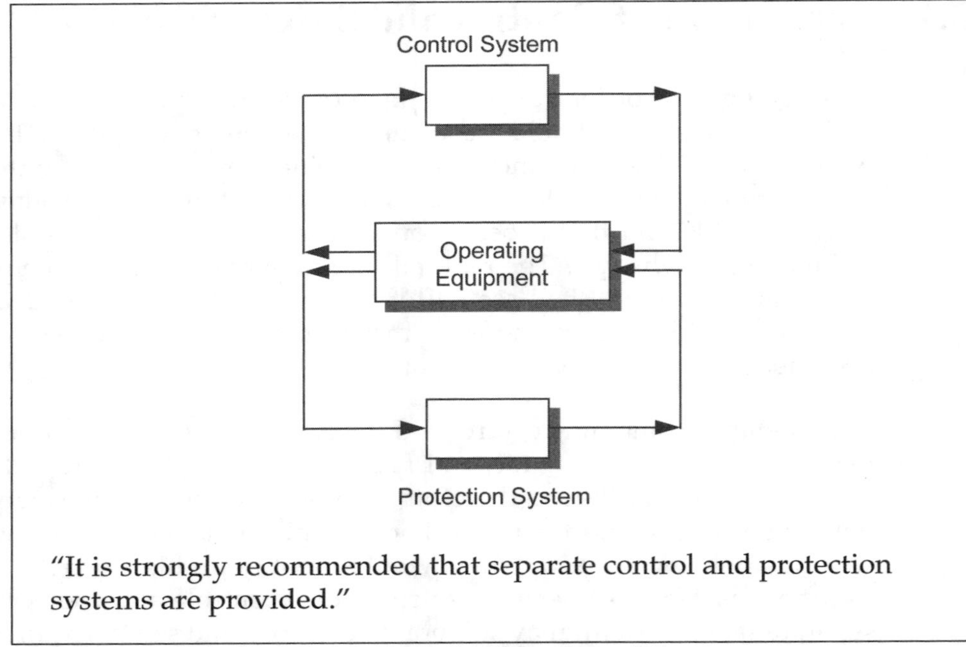

Figure 3-1: U.K. Health & Safety Executive Recommendations

3.4.2 AIChE - CCPS

The American Institute of Chemical Engineers (AIChE) text [Ref. 3] is a set of "guidelines." About 50 pages of the text are devoted to what are termed interlock systems with additional good background material. Being a guideline, the text does not use words that imply mandatory requirements such as "shall." Instead, there are statements like 'normally...' that are always open to interpretation. However, some statements, like the ones below, are very straightforward.

"Normally, the logic solver(s) are separated from similar components in the Basic Process Control System (BPCS). Furthermore, SIS input sensors and final control elements are generally separate from similar components in the BPCS." What basis could you put forth for not following what a textbook written by industry experts—all end users at that—considered 'normal' and 'generally' appropriate?

"Provide physical and functional separation and identification among the BPCS and SIS sensors, actuators, logic solvers, I/O modules, and chassis..." In other words, physically separate the components so a failure of one will not impact the other. Implement different logic, because if the same logic is performed in both systems and there is an error in the functional specification, both systems will have the same error. Identify the systems differently on drawings and out in the field so people realize that, "Oh, this is a safety device. I need to be careful and follow additional procedures."

3.4.3 IEC 61508

This international standard [Ref. 4] is for *all* industries and covers the use of relay, solid state, and programmable safety systems (or E/E/PES; electric, electronic, and programmable electronic systems).

"The EUC (equipment under control) control system shall be separate and independent from the E/E/PE safety-related systems, other technology safety-related systems and external risk reduction facilities."

If this requirement cannot be met then the entire control system needs to be designed as a safety-related system and follow the requirements laid out in the standard. One such example would be aircraft controls. It's difficult to imagine a separate safety system designed to make an airplane at 35,000 feet fail to its safe state. In some cases control and safety are so intermingled that the entire system must be designed as safety-related.

3.4.4 ANSI/ISA-84.00.01-2004

The original 1996 version of this standard [Ref. 5] had the two following statements regarding separation. These statements were very black and white and easy to interpret.

"Sensors for SIS shall be separated from the sensors for the BPCS." Two exceptions were allowed: if there was sufficient redundancy, or if a hazard analysis determined there were enough other safety layers to provide adequate protection.

"The logic solver shall be separated from the BPCS..." There were exceptions for cases where it was not possible to separate control from safety functions, such as high speed rotating equipment like gas turbines.

ISA standards are generally revised every five years. Rather than completely rewrite the ISA SP84's standard, the committee agreed to adopt the IEC 61511 standard. The wording in the IEC standard regarding separation is not as clear-cut as in the 1996 version of the ISA SP84's standard. In fact, an appeal on this very issue delayed the release of the new standard an entire year. Users simply do not want separation *mandated* on them. Many believe that for very small systems, or systems with low performance requirements, combining control and safety may be acceptable, assuming things are still done properly. The new standard states the following:

Clause 9.5.1 states, "The design…shall be assessed to ensure that the likelihood of common cause…and dependent failures between… protection layers and the BPCS are sufficiently low in comparison to the overall safety integrity requirements of the protection layers." The definition of 'sufficiently low' is naturally open to interpretation, as well as what constitutes a proper assessment.

Clause 9.4.3 states, "If a risk reduction greater than 10 is claimed for the BPCS, it shall be designed to the standard." Performance terms such as "risk reduction" will be described later in this text. What this clause means is that if you are going to combine control and safety in one box and claim a level of performance even in the lowest SIL 1 range, the system will need to be designed according to the standard. That might be feasible for a small system, but would be totally impractical for a large system.

Clause 11.2.4 states, "If it is intended not to qualify the basic process control system to this standard, then the basic process control system shall be designed to be separate and independent to the extent that the functional integrity of the safety instrumented system is not compromised." Again, statements such as "to the extent that…" are always open to interpretation.

Clause 11.2.10 states, "A device used to perform part of a safety instrumented function shall not be used for basic process control purposes, where a failure of that device results in a failure of the basic process control function which causes a demand on the safety instrumented function, unless an analysis has been carried out to confirm that the overall risk is acceptable." In other words, even field devices should be separate. However, there is still an "out" if "an analysis" can justify combining systems. What sort of justification? Simply stated, one that you would be willing to use in court.

3.4.5 API RP 14C

The American Petroleum Institute, Recommended Practice 14C "Analysis, Design, Installation & Testing of Basic Surface Safety Systems for Offshore Production Platforms [Ref. 6]" applies to the design of offshore platform shutdown systems. It prescribes how to determine the inputs and outputs for the shutdown system for different process units. It says nothing, however, on the design of the logic box. (It assumes most logic systems are pneumatic.)

"The safety system should provide two levels of protection... The two levels of protection should be independent of and in addition to the control devices used in normal operation."

It's surprising how open to interpretation some of these documents are. For example, would a relief valve and a gas detection system be considered "two levels of protection independent of and in addition to the control devices"? 14C tells you how to determine I/O and functional logic for the system. It doesn't tell you what sort of logic box to use. People have installed pneumatic, relay, PLC, and Triple Modular Redundant (TMR) systems on offshore platforms and *all* claim to have met RP14C, and indeed they may all have.

3.4.6 API RP 554

The American Petroleum Institute, Recommended Practice 554 "Process Instrumentation and Control [Ref. 7]" applies to refineries. Its statements regarding separation are rather straightforward. "The shutdown and control functions should be in separate and independent hardware... ," and "Shutdown devices should be dedicated and separate from pre-shutdown alarm and the process control system." In other words, safety-related logic systems and field devices should be dedicated and separate from control and alarm systems. Note that this document, as well as 14C, is a recommended practice, not a mandatory standard. It uses the word "should" rather than "shall".

3.4.7 NFPA 85

The National Fire Protection Association has written standards that apply to boiler and burner management systems, such as "Boiler & Combustion Systems Hazards Code [Ref. 8]." These might be viewed as examples of 'prescriptive' standards. Prescriptive standards tend to have black and white statements that are easy to interpret and, therefore, comforting to some. This standard makes a very clear call for separation. "Requirement for Independence: The logic system performing the safety functions for burner management shall not be combined with any other logic system." A DCS might be utilized for the 6,000 I/O associated with a huge industrial boiler, but the burner management system shall be a separate system. There is another statement that says, "The logic system shall be limited to one boiler only." In other words, using a redundant logic system capable of multiple programs to handle multiple burner management systems is not acceptable. However, facilities that are self-insured may not follow the NFPA standard.

3.4.8 IEEE 603

The Institute of Electric and Electronic Engineers has written standards that apply to the design of nuclear generating stations, such as "Standard Criteria for Safety Systems for Nuclear Power Generating Stations." [Ref. 9] The process industry is obviously not as restrictive as the nuclear industry, but the concept of separate safety layers is rather clear cut here.

"The safety system design shall be such that credible failures in and consequential actions by other systems shall not prevent the safety system from meeting the requirements."

Some nuclear facilities use a quad redundant computer arrangement, dual in series with dual in parallel. And as if that weren't enough, some used *two* of these quad redundant systems together! The U.S. typically uses quad systems backed up with 'conventional analog' systems.

An article in an English journal a few years back described the software testing for the programmable safety system of an English nuclear power station. It took 50 engineers 6 months to proof test the system! It's doubtful that many companies in the petrochemical industry could afford 25 man/years to proof test each of their safety systems! Someone with the Nuclear Regulatory Commission (NRC) told one of the authors that it wasn't 25 man-years, but more like several *hundred* man-years!

3.5 Common Cause and Systematic/Functional Failures

There are a variety of names and definitions for common cause. In essence, a common cause failure can be defined as a single stressor or failure that impacts multiple items or portions of a system.

One way of referring to such problems is the "Beta factor." This is the percentage of all identified failures in one "leg" or "slice" of a *redundant* system that might impact identical components and make the entire system fail. For example, suppose a central processor unit (CPU) has a failure rate of 10E-6 failures/hour. If the CPU were triplicated and all three were placed side by side in the same chassis, certain failures might impact two or three CPUs at once (e.g., a design, manufacturing, or software error). A Beta factor of 10% would mean that the system would act in a non-redundant manner with a failure rate of 1E-6 failures/hour. Note that the Beta factor model is purely empirical. It's not reality, just a model or estimation of reality. There are a number of documented techniques for estimating Beta value percentages. [Ref. 4 and 10]

Systematic failures, also called functional failures, impact an entire system and are therefore sometimes categorized as common cause failures. Examples of systematic failures usually include human errors in design, operation, and maintenance. However, heat, vibration, and other external factors can still be accounted for in the definition. Such failures may impact a redundant or non-redundant system. A systematic error in specifying a solenoid valve could result in the valve not operating properly at low temperatures. If redundant valves were used, the systematic failure would also be a common cause failure. While some systematic failures may be relatively easy to identify (e.g., a design error), the exact number or percentage (if accounting for them using the Beta factor) of such failures is usually more difficult to quantify. How accurately can one predict, for example, how often an engineering design error might occur, or that a maintenance technician might calibrate all three transmitters incorrectly?

One study in the nuclear industry found that 25% of all failures in nuclear power stations were related to common cause. [Ref. 10] Investigations in redundant control systems have found common cause percentages between 10% and 30%. It's surprising how some very redundant systems still have single points of failure. One networked computer systems had 7-way redundant communications. Unfortunately, they all ran through the same fiber optic cable which, when cut, disabled the entire system. [Ref. 11] Obviously no one would *intentionally* design a system this way. The point is, things eventually grow to the point where one individual can't "see" everything and certain items inevitably fall through cracks.

If control and safety functions are performed in the same system, there will always be potential for common cause faults. The more the systems are physically separated, the more unlikely it will be that single failures can affect them both. A simple phrase sums this thought up fairly well, "Don't put all your eggs in one basket." No matter how sturdy or reliable the basket may be, there will always be some unforeseen circumstances where it will drop. Everything fails, it's just a matter of when.

3.5.1 Human Issues

The various standards state that control and safety systems should be separate, but they don't really state *why*. The issues are actually rather straightforward. Much of it boils down to human issues.

Safety systems require strict management of change. If one were to combine control and safety in the same system, how effectively can you grant operators access to certain functions they need to change frequently, but lock them out of other functions that they should not? How assured can you be that changes to one area of the system (e.g., software logic) do not impact a safety function elsewhere in the system? How assured can you be that placing a control loop in bypass, which is common practice, will not impact the operation of a safety alarm or action? The only way to have any real assurance is to have separate systems.

An actual example might help put this in perspective. This story was relayed to one of the authors during a lunch at one of the ISA SP84 committee meetings. A user company acquired an existing facility from another organization. An engineer from the parent company was sent to audit the control and safety systems at the acquired plant. The engineer found that all control and safety functions were performed in the DCS. Further checks were performed. It was found that one third of the safety functions had simply vanished (i.e., been deleted). Another third were found in bypass. The remaining third were tested and found not to be functional. One might make the argument that establishing and following proper procedures would have prevented such a situation. After all, procedures must be established for bypassing, maintaining, testing and management of change of safety systems. Unfortunately, procedures will often be violated, whether on purpose or by accident. After all, if we were all proper drivers and followed proper procedures, there would be no need for seat belts or air bags.

Summary

Process control systems are active and dynamic, therefore most faults are inherently self-revealing. Safety systems are passive or dormant, therefore many faults are *not* self-revealing. Safety systems require either manual testing or effective self-diagnostics, something many general purpose control systems do not incorporate very effectively.

Control systems are designed to allow relatively easy access so operators can make the frequent changes that are required. Safety systems, however, require strict security procedures and access control in order to prevent inadvertent changes. The control system, or any changes made to it, must not prevent the safety system from functioning properly.

In the not-too-distant past, process control and safety systems were implemented in separate, diverse technology systems. Pneumatic control systems are being replaced by software-based DCSs. Relay-based safety systems are being replaced by software-based PLCs. Incorporating both in one combined system offers potential benefits of a single source of supply, simplified stores, maintenance and training, and possibly lower cost. Standards, guidelines, and recommended practices from numerous industries, however, all *strongly* discourage such a practice. The issue has little to do with the reliability of modern control systems.

The design of safety systems requires following strict and conservative design requirements. Analysis, documentation, design, operation, maintenance, and management of change procedures all require extra effort for safety systems. Implementing both control and safety in one system means that the extra effort must be applied to the *entire* control system. This is simply not economically feasible for any reasonably sized system. Placing all your eggs in one basket is rarely a good idea. Everything fails, it's just a matter of when.

If an organization chooses to diverge from the standards, they better have a valid, documented reason for doing so. A simple thought to keep in the back of your mind is, "How will I justify this decision in court?" If anything were to ever happen, that's what it may come down to. If your answer for violating the standards is, "It was cheaper that way", the court may not respond favorably.

References

1. Kletz, Trevor A. *Computer Control and Human Error.* Gulf Publishing Co., 1995.
2. *Programmable Electronic Systems in Safety Related Applications, Part 1 - An Introductory Guide.,* U.K. Health & Safety Executive, 1987.
3. *Guidelines for Safe Automation of Chemical Processes.* American Institute of Chemical Engineers - Center for Chemical Process Safety, 1993.
4. IEC 61508-1998. *Functional Safety of Electrical/Electronic/Programmable Electronic Safety-Related Systems.*
5. ANSI/ISA-84.00.01-2004, Parts 1-3 (IEC 61511-1 to 3 Mod). *Functional Safety: Safety Instrumented Systems for the Process Industry Sector* and ISA-84.01-1996. *Application of Safety Instrumented Systems for the Process Industries.*
6. API RP 14C-2001. *Analysis, Design, Installation, and Testing of Basic Surface Safety Systems for Offshore Production Platforms.*
7. API RP 554-1995. *Process Instrumentation and Control.*
8. NFPA 85-2001. *Boiler and Combustion Systems Hazards Code.*
9. IEEE 603-1991. *Standard Criteria for Safety Systems for Nuclear Power Generating Stations.*

10. Smith, David J. *Reliability, Maintainability and Risk: Practical Methods for Engineers.* 5th Edition. Butterworth-Heinemann, 1997.

11. Neumann, Peter. G. *Computer Related Risks.* Addison-Wesley, 1995.

12. ANSI/ISA-91.00.01-2001. *Identification of Emergency Shutdown Systems that are Critical to Maintaining Safety in Process Industries.*

4
PROTECTION LAYERS

Chapter Highlights

Accidents rarely have a single cause. Accidents are usually a combination of rare events that people initially assumed were independent and would not happen at the same time. Take as an example the worst chemical accident to date, Bhopal, India, where an estimated 3,000 people died and 200,000 were injured. [Ref. 2]

The material that leaked in Bhopal was MIC (methyl isocyanate). The release occurred from a storage tank which held more material than allowed by company safety requirements. Operating procedures specified using the refrigerant system of the storage tank to keep the temperature of the material below 5 degrees C. A temperature alarm would sound at 11 degrees. The refrigeration unit was turned off due to financial constraints and the material was usually stored at nearly 20 degrees. The temperature alarm threshold was changed from 11 to 20 degrees.

A worker was tasked to wash out some pipes and filters which were clogged. Blind flanges were not installed as required. Water leaked past valves into the tank containing MIC. Temperature and pressure gauges which indicated abnormal conditions were ignored because they were

believed to be inaccurate. A vent scrubber, which could have neutralized the release, was not kept operational because it was presumed not to be necessary when production was suspended (as it was at the time). The vent scrubber was inadequate to handle the size of the release anyway. The flare tower, which could have burned off some of the material, was out of service for maintenance. It was also not designed to handle the size of the release. Material could have been vented to nearby tanks, but gauges erroneously showed them to be partially filled. A water curtain was available to neutralize a release, but the MIC was vented from a stack 108 feet above the ground, too high for the water curtain to reach. Workers became aware of the release due to the irritation of their eyes and throats. Their complaints to management at the time were ignored.

Workers panicked and fled, ignoring four buses that were intended to be used to evacuate employees and nearby residents. The MIC supervisor could not find his oxygen mask and broke his leg climbing over the boundary fence. When the plant manager was later informed of the accident, he said in disbelief, "The gas leak just can't be from my plant. The plant is shut down. Our technology just can't go wrong, we just can't have leaks."

Investigations of industrial accidents have found that a large number occurred during an interruption of production while an operator was trying to maintain or restart production. In each case the dangerous situation was created by a desire to save time and ease operations. In each case, the company's safety rules were violated. [Ref. 2]

The best and most redundant safety layers can be defeated by poor or conflicting management practices. Numerous examples have been documented in the chemical industry. [Ref. 1] One accident in a polymer processing plant occurred after operations bypassed all alarms and interlocks in order to increase production by 5%. In another, interlocks and alarms failed—at a normal rate—but this was not known because management had decided to eliminate regular maintenance checks of the safety instrumentation.

James Reason [Ref. 3] has described how organizational accidents happen when multiple safety layers fail. Figure 4-1a shows the design intent of multiple layers. If all the layers are effective (i.e., solid and strong), a failure will not propagate through them. However, in reality, the layers are not solid. They're more like Swiss cheese. The holes are caused by flaws due to management, engineering, operations, maintenance, and other errors. Not only are there holes in each layer, the holes are constantly moving, growing, and shrinking, as well as appearing and disappearing. It's now easy to visualize how, if the holes line up properly (Figure 4-1b), a failure can easily propagate through all of them.

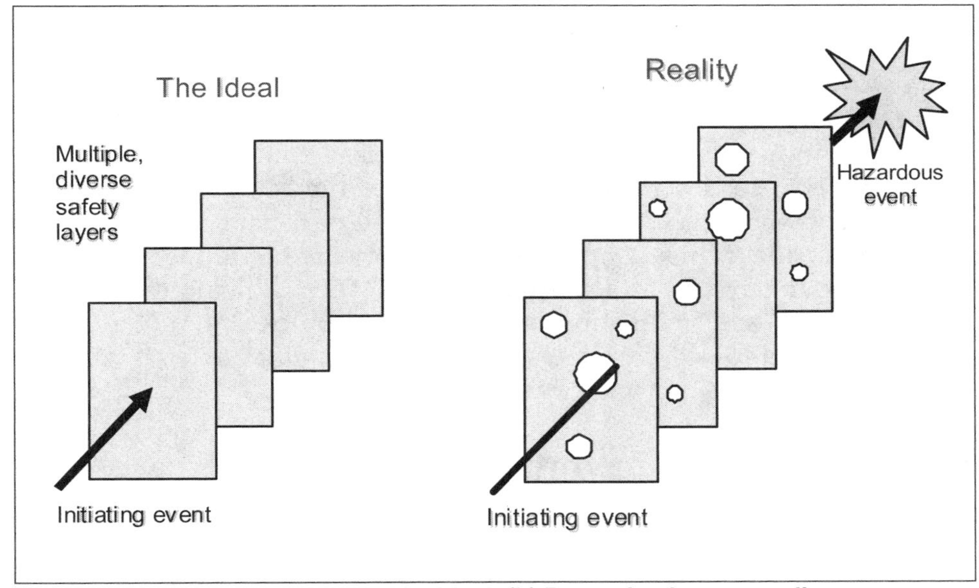

Figure 4-1: (a) Protection layers - the ideal; (b) Protection layers - reality

Figure 4-2, often referred to as "the onion diagram," appears in a number of different formats in most safety documents. It shows how there are various safety layers, some of which are prevention layers, others which are mitigation layers. The basic concept is simple: "don't put all your eggs in one basket." Some refer to this as "defense in depth."

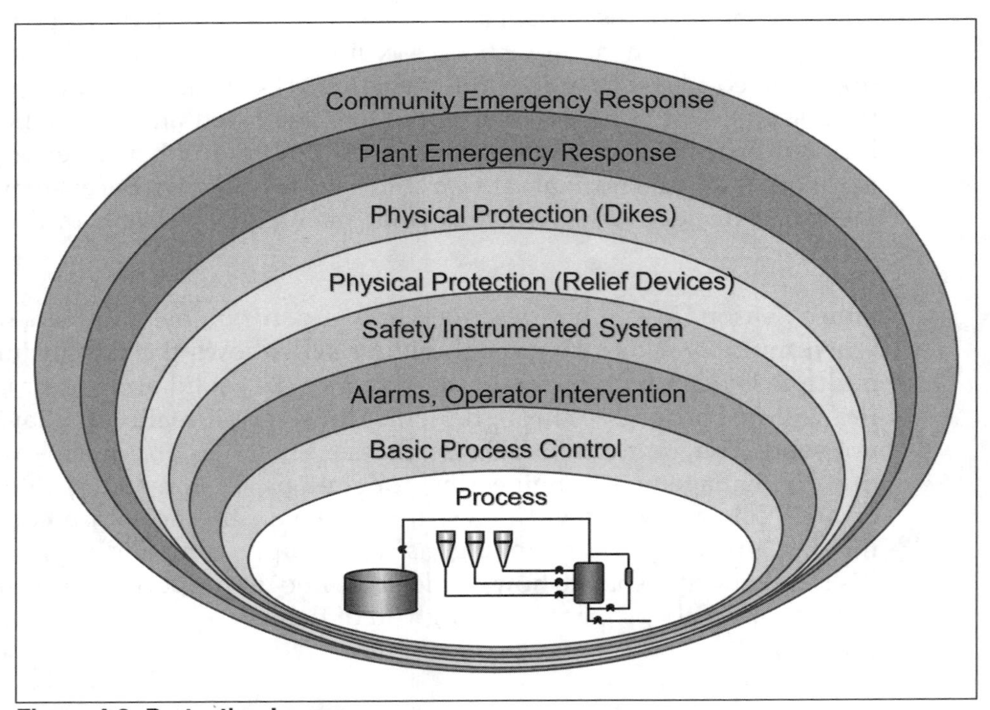

Figure 4-2: Protection Layers

Risk is a function of the probability (or frequency, or likelihood) of an event and its severity (or consequences). Multiple safety layers in any facility are designed to reduce one or the other. Prevention layers are implemented to reduce the probability of a hazardous event from ever occurring. Mitigation layers are implemented to reduce the consequences once the event has already happened. The following discussions on prevention and mitigation layers are examples only. The listing is not intended to be viewed as all possible layers that may be implemented or should be used in any one facility. [Ref. 5]

4.1 Prevention Layers

Prevention layers are implemented to reduce the probability or likelihood of a hazardous event from ever occurring. Multiple, diverse layers are generally used.

4.1.1 Process Plant Design

The process plant itself must be designed with safety in mind. This is why HAZOP (HAZard and OPerability studies) and other safety reviews, such as fault trees, checklists, what-if, etc., are performed.

A major goal within the process industry is to design inherently safe plants. Don't design a dangerous plant with the intention of throwing on lots of Band-Aids hoping they'll overcome the problems. Design it so the Band-Aids aren't even necessary. Work with low pressure designs, low inventories, non-hazardous materials, etc. As Trevor Kletz said, "What you don't have can't leak."

It's surprising how much rebellion there is to this, along with most other sorts of change. The typical complaint is, "But we can't *afford* to do all that!" Information one of the authors has seen in various industries would indicate that inherently safer systems *may* have a higher initial cost (although not always), but they offer a *lower* cost of ownership *over the life* of any project. The same applies to quality management. Think about how many companies said they couldn't afford that, either. Now, you can't afford *not* to have it.

Eliminating or reducing hazards often results in *simpler* designs, which may in itself reduce risk. The alternative is to add protective equipment to control hazards, which usually adds complexity.

One example of the rebellion against inherently safer designs can be found in everyday life. The Refrigerator Safety Act was passed because children were being trapped and suffocated while playing in unused refrigerators. Manufacturers *insisted* they could not afford to design safer latches. When *forced* to do so, they introduced simple, magnetic latches. These permitted the door to be opened from the inside, thus *eliminating* the hazard. The newer design was also *cheaper* than the older one! [Ref. 2]

4.1.2 Process Control System

The process control system is the next layer of safety. It controls the plant for optimum fuel usage, product quality, etc. It attempts to keep all variables (e.g., pressure, temperature, level, flow, etc.) within safe bounds, therefore it can be considered a safety layer. However, a control system failure may also initiate a hazardous event.

Automation usually does not eliminate humans from the system. In fact automation frequently raises human tasks to new levels of complexity. If computers are being used to make more and more decisions because human judgment and intuition are not satisfactory, then it may be an error to have a human act as the final arbiter. Experience has shown that humans make poor monitors of automated systems. Tasks that require little active operator action may result in lowered alertness and vigilance, and can lead to complacency and over-reliance on automated systems. Long periods of passive monitoring can make operators unprepared to act in emergencies. Some have commented that "computer control turns operators into morons." One way to solve this problem would be to involve operators in safety analyses and design decisions up front and throughout development. Involve operators more, not less.

4.1.3 Alarm Systems

If the process control system fails to perform its function (for any number of reasons, such as an internal failure or a loop being placed in bypass) alarms may be used to alert the operators that some form of intervention is required on their part.

Alarm and Monitoring Systems Should:

1. Detect problems as soon as possible, at a level low enough to ensure action can be taken before hazardous conditions are reached.
2. Be independent of the devices they're monitoring (i.e., they should not fail if the system they're monitoring fails).
3. Add as little complexity as possible.
4. Be easy to maintain, check, and calibrate.

Alarm and monitoring systems are considered to be the safety layer where people get actively involved. Operators will generally be required in plants for the simple reason that not *everything* can be automated. It is essentially impossible for designers to anticipate *every* possible set of conditions that might occur. Human operators may need to be considered since only they will be flexible and adaptable enough in certain situations.

This is a double-edged sword, however, because events not considered in the design stage will no doubt also not be included in operator training either. On the other hand, simply blindly following procedures has resulted in accidents. Deviation from the rules is a hallmark of experienced people, but it's bound to lead to occasional human error and related blame after the fact.

4.1.3.1 Human Reliability

One of the authors has heard some people say they did not want automated systems in their plants controlling safety. They wanted to rely on people who were educated and trained on the operation and dynamics of their process. While this may be acceptable for normal routine operations (although some might still argue against this), it is *not* the recommended scheme for critical emergency situations.

For example, accidents have occurred because:

1. Operators did not believe rare events were real or genuine.
2. Operators were overload with information, and failed to act.

People have been, and will continue to be, directly responsible for some accidents. Some in the industry, most notably Trevor Kletz, have done an excellent job documenting such cases. [Ref. 1] Hopefully, the rest of the industry will learn from these examples and not repeat them. However, even Kletz has said, "Accidents are not due to lack of knowledge, but failure to use the knowledge we already have." Unfortunately, history has shown that many of the accidents recur.

For example, there have been cases where the operators saw the alarm, knew what it meant, and *still* took no action. Either the alarm was considered a nuisance alarm ("Oh we see that all the time"—this was one of many problems at Bhopal) or they waited to see if anything else would happen (sometimes with catastrophic results).

When things do go wrong, they tend to cascade and escalate. One of the authors knows of one plant where there was a shutdown and the DCS printed out *17,000* alarm messages! Overwhelming the operators with this much information is obviously detrimental. *Too* much information is *not* a good thing. Kletz has documented other similar cases.

When faced with life threatening situations requiring decisions within one minute, people tend to make the wrong decisions 99% of the time. This was determined from actual studies done by the military. In other words, during emergencies, people are about the *worst* thing to rely on, no matter how well trained they may be.

4.1.4 Procedures

Some *might* consider operating and maintenance procedures as protection layers. This is a rather controversial subject. Inspections to detect corrosion and degradation of a vessel may help prevent accidents. Procedures limiting the operation of a unit to below its safety limits may help prevent accidents. Preventative maintenance to replace components before they fail may also help avoid accidents. However, all procedures will be violated at some point (intentionally or not). Also, with the push for cost and manpower reductions, procedures that may have been possible in the past may no longer be feasible today or in the future.

If procedures *are* to be accounted for as a protection layer, they need to be documented, people need to be trained to follow them, and their use must be audited. Engineers and managers that assume certain procedures are being followed may be appalled when operators and technicians inform them of what actually goes on in the plant. Then again, operators and technicians may actually be reluctant to inform others of these issues knowing full well it may only make their lives and tasks more difficult.

4.1.5 Shutdown/Interlock/Instrumented Systems (Safety Instrumented Systems – SIS)

If the control system and the operators fail to act, automatic shutdown systems take action. These systems are usually completely separate, with their own sensors, logic system, and final elements. (Please refer back to Chapter 3 on the separation issue.) These systems are designed to:

1. Permit a process to move forward in a safe manner when specified conditions allow, or
2. Automatically take a process to a safe state when specified conditions are violated, or
3. Take action to mitigate the consequences of an industrial hazard.

A safety system therefore can provide permissive, preventive, and/or mitigation functions. It's important to distinguish between a SIF (safety instrumented function) and a SIS (safety instrumented system). A SIF refers to a single function (e.g., high pressure shutdown, low level shutdown, etc.), whereas a SIS refers to all the combined functions that make up the overall system. Most SIFs contain a single input sensor and a single, or possibly a few, final elements. A SIS, however, may contain dozens, hundreds, or even thousands of inputs and outputs.

These systems require a higher degree of security to prevent inadvertent changes and tampering, as well as a greater level of fault diagnostics. The focus of this book is on these systems.

4.1.6 Physical Protection

Relief valves and rupture discs are one means of physical protection that could be used to prevent an overpressure condition. While this may serve to prevent a pressure vessel from exploding, venting of material may result in a secondary hazardous event (e.g., release of a toxic material) or fines due to an environmental violation.

4.2 Mitigation Layers

Mitigation layers are implemented to lessen the severity or consequences of a hazardous event once it has already occurred. They may contain, disperse, or neutralize the release.

4.2.1 Containment Systems

If an atmospheric storage tank were to burst, dikes could be used to contain the release. However, holding process fluids within dikes may introduce secondary hazards. Reactors in nuclear power plants are usually housed in containment buildings to help prevent accidental releases. The Soviet reactor at Chernobyl did not have a containment building, whereas the U.S. reactor at Three Mile Island did.

4.2.2 Scrubbers and Flares

Scrubbers are designed to neutralize a release. Flares are designed to burn off excess material. Note that Bhopal had both systems, yet both were not functioning during the maintenance phase the plant was in at the time.

4.2.3 Fire and Gas (F&G) Systems

A fire and gas system (F&G) is composed of sensors, logic solver, and final elements designed to detect combustible gas, toxic gas, or fire and either a) alarm the condition, b) take the process to a safe state, or c) take action to mitigate the consequences of a hazardous event. Sensors may consist of heat, smoke, flame, and/or gas detectors, along with manual call boxes. Logic systems may consist of conventional PLCs, distributed control systems (DCS), safety PLCs, special purpose PLCs designed specifically for F&G applications, or dedicated multi-loop F&G controllers. Final elements may consist of flashing/strobe lights, sirens, a telephone notification system, exploding squibs, deluge systems, suppression systems, and/or process shutdowns.

A gas detection system doesn't *prevent* a gas release; it merely indicates when and where one has already occurred. Similarly, a fire detection system doesn't *prevent* a fire; it merely indicates when and where one has already occurred. These are traditional mitigation layers that are used to lessen the consequences of an event that has already happened. In the U.S. these are often "alarm only" systems—they do not take any automatic control actions, and the fire crews must go out and manually put out the fire. Outside the U.S. these systems frequently take some form of automated action or may be integrated with the shutdown system.

One major difference between shutdown and fire and gas systems is that shutdown systems are normally energized (i.e., de-energize to shutdown), whereas fire and gas systems are normally de-energized (i.e., energize to take action). The reasoning for this is actually rather simple. Shutdown systems are designed to bring the plant to a safe state, which usually means stopping production. Nuisance trips (i.e., shutting the plant down when nothing is actually wrong) are economically detrimental due to lost production downtime, but nuisance trips are generally not catastrophic in terms of safety. Actually, studies have shown that while shutdown/startup operations only account for about 4% of the total operating time, about 25% of all accidents happen during that 4%. [Ref. 3] Fire and gas systems are designed to protect equipment and people. Spurious operations of these systems can damage certain pieces of equipment and possibly even result in deaths (e.g., an unannounced Halon or CO_2 dump in a control room). If systems are normally de-energized, nuisance trip failures are highly unlikely.

Debate continues as to whether a fire and gas system is, or is not, a safety instrumented system. The answer to this question should not be taken lightly, because it has an impact on the cost, hardware, applicable standards, responsibilities, design criteria, operation, and maintenance of the system.

4.2.4 Evacuation Procedures

In the event of a catastrophic release, evacuation procedures can be used to evacuate plant personnel and/or the outside community from the area. While these are procedures only and not a physical system (apart from sirens), they may still be considered one of the overall safety layers.

A siren was used at Bhopal to warn citizens nearby. However, it had the unintended effect of *attracting* nearby residents (much like a moth to a light) rather than repelling them!

4.3 Diversification

Investors understand the need for diversification. If an investor places his entire life savings in one stock account, and that account goes bust, the investor is out his entire life savings. It's safer to spread the risk among multiple stock accounts. That way, if any one account were to fail, the others would hopefully protect against total loss. The same applies to safety layers in process plants. Don't put all your eggs in one basket, no matter how good the basket may be. Everything fails, it's just a matter of when. In general, the more layers there are, and the more diverse they are, the better. In addition, each layer should be as simple as possible. Failure of one layer should not prevent another layer from performing its intended function.

The benefit of multiple, diverse layers can be seen in Figure 4-3. This figure is also useful for understanding the concept of Layer of Protection Analysis (LOPA) that will be covered in Chapter 6. Imagine the frequency of a hazardous event that might cause multiple fatalities at once per year due to the basic process design. This is represented in the diagram by the vertical line on the right labeled "risk inherent in the process". No one would consider it "tolerable" to have a catastrophic event once per year at every process facility that resulted in multiple fatalities! A more desirable goal might be something in the range of 1/100,000 per year. This is represented in the diagram by the vertical line on the left labeled "tolerable risk level". Each safety layer serves to minimize the risk. However, no layer is perfect, as shown in Figure 4-1b. How effective (or strong) each layer is can be represented graphically by how long (or wide) each arrow is. In other words, what is the "risk reduction factor" of each safety layer? The larger the risk reduction, the larger the arrow. No single layer can lower the risk by a factor of 100,000. In other words, no single layer offers a level of performance that could lower the hypothetical yearly event down to 1/100,000 per year. However, five independent layers each offering a risk reduction factor of 10 could.

These different layers were historically implemented in totally separate systems supplied by different vendors. They often utilized very different technologies. This level of diversification made common cause failures (i.e., a single failure that affects multiple components) extremely unlikely. However, many of these systems today utilize computers and software. What might happen if one were to combine fire and gas detection, shutdown, and alarm functionality all within the process control system? Doing more in one box doesn't make it perform any better. Running more programs in a computer doesn't make it run any better or faster. In fact, the opposite is usually the case. A Volkswagen Beetle was designed to carry a certain number of people a certain distance at a certain speed. Cramming 10 fraternity brothers into it won't make it go any faster. Similarly, doing more in the control system won't make it perform any better. In other words, the size of the Basic Process Control System (BPCS) arrow will not change at all. In fact, the IEC standards set a claim limit on the BPCS at a risk reduction factor of 10. One can now easily see that combining too many functions in the control system may in fact *lower* the overall safety of a facility, not improve it through the use of new technologies!

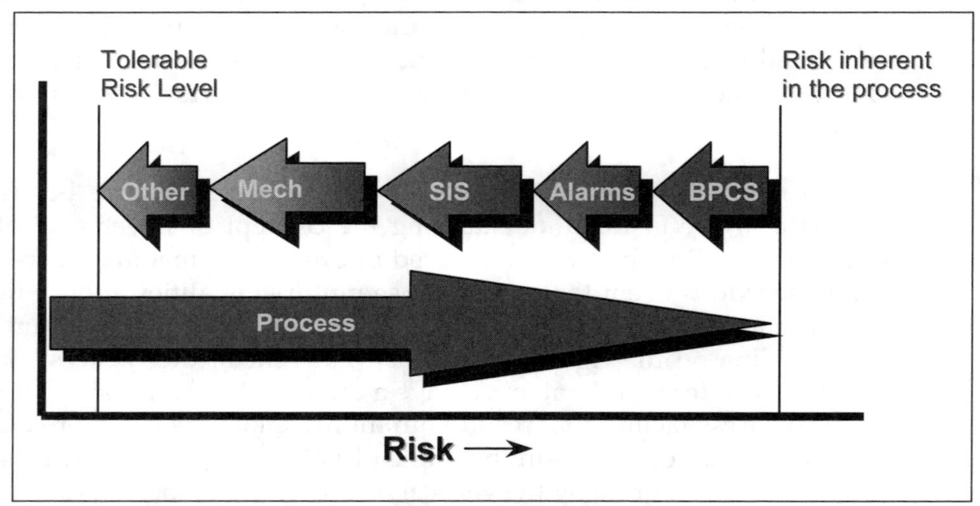

Figure 4-3: Reducing risk with multiple protection layers

Summary

Accidents are usually a combination of rare events that people initially assumed were independent and would not happen at the same time. One method of protecting against such instances is to implement multiple, diverse safety layers. This makes it harder for any one initiating event to propagate through all the layers culminating in a hazardous event.

One should implement multiple, diverse, independent safety layers. Some layers are in place to prevent accidents from occurring (prevention layers). Others are in place to lessen the consequences of an event once it has already happened (mitigation layers).

In general, the more safety layers, the better. Combining too much functionality in any one single layer may actually have the detrimental effect of degrading safety, not improving it. However, the best defense is to remove the hazard during the initial design phase. "What you don't have can't leak." Inherently safe designs can result in simpler designs with lower overall costs.

References

1. Kletz, Trevor A. *Lessons From Disaster: How Organizations Have no Memory and Accidents Recur.* Gulf Publishing, 1993.
2. Leveson, Nancy G. *Safeware - System Safety and Computer.* Addison-Wesley, 1995.
3. Reason, J. T. *Managing the Risks of Organizational Accidents.* Ashgate, 1997.
4. *Oil and Gas Journal.* Aug. 27, 1990.
5. *Guidelines for Safe Automation of Chemical Processes.* American Institute of Chemical Engineers - Center for Chemical Process Safety, 1993.

5

DEVELOPING THE SAFETY REQUIREMENT SPECIFICATIONS

Chapter Highlights

Nothing can go wrong (click), go wrong (click), go wrong (click)...

5.1 Introduction

Once the need for a safety instrumented system (SIS) is identified and the target safety integrity levels (SIL) have been established for each safety instrumented function (SIF), the safety requirement specification (SRS) needs to be developed. The SRS can either be a single document or a series of documents. The SRS lists the requirements for all functions to be performed by the system. This should also include the application software requirements. The SRS consists of two main parts:

1. The functional requirement specifications
2. The integrity requirement specifications

The functional requirements describe the *logic* of the system. In other words, it defines *what* each safety function should do. For example, low pressure in vessel A will cause valve B to close. The integrity requirements describe the *performance* needed for each function. In other words, it defines the *capability* or *how well* each safety function should perform. For example, the probability or likelihood that valve B will close when the pressure in vessel A is low needs to be greater than 99%. When the term "specification" is used hereafter in this chapter, it refers to both the functional requirement and the integrity requirement specifications.

It's best to design plants that are inherently safe, ones where hazards won't even exist. Unfortunately, this is not always possible. There are times when the only way to prevent a hazard is to add instrumentation that will monitor the process conditions and operate devices to take necessary actions. Deciding what process parameters are to be monitored, what the safety actions should be, and how well the system needs to operate, are all reflected in the functional and integrity specifications.

Chapter 2 noted that the English Health and Safety Executive examined 34 accidents that were the direct result of control and safety system failures in a variety of different industries. Their findings, which are summarized in Figure 2-1, indicated that 44% of accidents were due to incorrect specifications (both the functional as well as the integrity specification). The second highest percentage of failures (20%) were due to changes made after commissioning. The importance of ensuring that the system specifications are correct cannot be overemphasized. Poor specifications will result in systematic (design) failures. In other words, the system may do what it was designed to do, but that action may be incorrect. Systematic failures are non-random and usually occur under the same conditions. These failures cannot be adequately quantified and are generally controlled by ensuring that the all lifecycle activities are followed.

This chapter identifies some of the problems associated with specification development, reviews ANSI/ISA-84.00.01-2004 (IEC 61511 Mod) requirements, and includes examples of typical documentation that needs to be developed.

5.2 Why Are 44% of Accidents Caused by Incorrect Specifications?

In order to understand *why* the majority of accidents were caused by incorrect specifications, it's necessary to focus on the systems and procedures used to *produce* the specifications. As per the life cycle model outlined in Chapter 2, the following activities need to occur *prior* to the preparation of the specification.

1. Conceptual process design
2. Hazard analysis and risk assessment
3. Application of non-SIS layers
4. SIL determination if an SIF is required

These four steps form the foundation of a correct specification. The procedures used to complete these steps within an organization must therefore be effective and adequate. If procedures are non-existent or inadequate, it doesn't matter how much time or effort is expended in the development of system specifications—the fundamental design basis for the system may end up being incorrect.

The purpose of this book is not to define process design procedures or hazard analysis techniques. Numerous publications on process design and hazard analysis are available from the American Institute of Chemical Engineers Center for Chemical Process Safety and other sources. It should be pointed out that identifying and assessing hazards is *not* what makes a process or system safe. It's the information obtained from the hazard analysis, and more importantly, *the actions taken*, that determine system safety.

Simply performing a HAZard and OPerability study (HAZOP) won't make your plant any safer. After all, the study simply produces a report or a piece of paper. Following the recommendations of the study is what makes things safer.

In addition to the systematic and procedural issues just mentioned, the following are additional reasons why such a large percentage of failures have been contributed to incorrect specifications:

- management systems
- procedures
- scheduling of assessment
- participation of key personnel
- responsibilities not well defined
- training and tools
- complexity and unrealistic expectations
- incomplete documentation
- inadequate final review of specification
- unauthorized deviation from specification

5.2.1 Management Systems

A management system (i.e., a system to ensure effectiveness) must be in place for the lifecycle model to be effective. A management system usually consists of the following steps:

1. identifying the activities in the lifecycle model, and the expected results of those activities
2. setting performance standards against which the results will be compared
3. measuring the actual performance
4. evaluating and comparing the actual performance against the set standards
5. taking necessary corrective actions based on the evaluation

Although it may be relatively easy to develop internal standards and procedures, it may be more difficult to develop and implement a management system to ensure full compliance. The root cause of many problems can often be traced back to nonexistent or ineffective management systems. ANSI/ISA-84.00.01-2004, Parts 1-3 (IEC 61511-1 to 3 Mod) highlights the requirements, importance and methodology for the management of functional safety.

F. Bird and G. Germain's *Practical Loss Control Leadership* is an excellent text on management systems. [Ref. 2] The term *management system,* although widely used, is somewhat of a misnomer. These systems often have little to do with "management." A better term commonly used is *managing systems.*

5.2.2 Procedures

Procedures need to be developed, reviewed, approved, and understood for all activities in order to effectively perform hazard analyses and SIL determinations. For example, are the SIL determination procedures intended to cover personnel safety only, or are they intended to cover other risk receptors (e.g., environment incidents, financial loss, equipment/property damage, business interruption, business liability, company image, lost market share)? This should be clearly reflected in the procedures.

Hazard analysis usually involves techniques such as fault tree analysis (FTA), failure mode and effects analysis (FMEA), hazard and operability studies (HAZOP), and other similar methods. The HAZOP technique is probably the most widely accepted and used method in the process industry.

SIL determination techniques (the topic of Chapter 6) are not as well established. Large organizations usually have the resources to develop the techniques and procedures. Smaller organizations may have to rely on outside consultants for assistance.

5.2.3 Scheduling of Assessment

Safety assessments are sometimes carried out without the method being well defined or understood simply due to the rush to have the activity completed. Those responsible for planning the assessments need to ensure that the process design and necessary documentation have been completed to such an extent that a meaningful safety analysis can be done. The assessment must not be a mere formality simply to satisfy project management schedules or procedures.

The SIL determination should be an extension of hazard review. This ensures that synergies and understandings developed during the safety assessment are extended to the SIL determination.

5.2.4 Participation of Key Personnel in the Review Process

Most units and systems in process plants are far too complex for any one person to be a total expert on them. One person simply can't be a master on the chemistry of a process, the materials used to contain it, the instrumentation used to monitor it, the control systems used to control it, the rotating equipment used to drive it, the electrical systems used to power it, and the safety techniques used to analyze it.

For this very reason, HAZOP and other hazard analysis techniques usually require a multi-discipline *team*. Persons representing the project, process, operations, safety, maintenance, instrumentation, and electrical groups are typically required. These studies analyze what might go wrong in the operation of a plant. If a hazard is identified and considered significant, a decision must be reached on how to prevent or mitigate it.

When proceeding through the different phases of the hazard analyses and SIL determinations, it's important that each phase not be regarded as a distinct activity. In other words, it should not be viewed that upon completion of one particular phase, the results are simply passed to another team over a "solid brick wall."

5.2.5 Responsibilities Not Well Defined

For each of the four activities listed at the beginning of Section 5.2, the individuals with the lead/responsible roles needs be identified and agreed upon. The individuals assigned must clearly understand their roles and ensure that the analyses are completed in accordance with established procedures. Typical lead roles for each of the four activities are as follows:

Activity	Lead role responsibility
Conceptual process design	Process design
Hazard analysis and risk assessment	Safety engineering
Application of non-SIS layers	Process design
SIL determination if an SIF is required	Safety engineering

5.2.6 Training and Tools

The teams involved with the assessments need to be adequately trained and provided with the necessary tools. For example, prior to an assessment it may be necessary that process control and electrical personnel receive process training, or that process personnel are adequately trained on the type and capabilities of the control systems. This is essential in order to ensure proper dialog and understanding during the review process. The capabilities and training needs of each team member should be identified prior to the analyses.

5.2.7 Complexity and Unrealistic Expectations

Generally speaking, the simpler a design the better. Complex designs are more difficult to understand, visualize, and review. Take as an example one accident described by Trevor Kletz. [Ref. 1]

An accident in one plant involved a highly hazardous ethylene oxide process. Three valves in series were installed, with two bleed valves between them. As if that weren't enough, these five valves were duplicated so the system could be tested while the plant was on-line without interrupting production.

Operation of the system was carried out automatically by a computer. As a result of a software error, the computer momentarily operated the valves incorrectly. The ensuing explosion and fire of three tons of gas caused extensive damage and resulted in lost production for several months. The software error could be attributed to complexity.

If a safety system is required to prevent or mitigate a hazard, requirements should be stated as simply as possible. Team personnel should refrain from establishing solutions that are too complex, as this can lead to unrealistic and unwarranted expectations.

5.2.8 Incomplete Documentation

Process control and/or instrumentation personnel are usually responsible for the preparation of the safety requirement specifications. The method for documenting the specifications needs to be simple enough for all individuals to understand what is to be provided.

Table 5-1 summarizes the information that should be included in the specifications. The "details of requirement" column identifies whether information pertaining to each item should be provided for the particular application. Comments should also be included as necessary. Additional items are also summarized in the design checklist included in Chapter 15.

Table 5-1: Summary information that should be included in the safety requirements specifications

Item	Details of Requirement
Documentation and Input Requirements	
P&IDs	
Cause and effect diagram	
Logic diagrams	
Process data sheets	
Process information (incident cause, dynamics, final elements, etc.) of each potential hazardous event that requires an SIF	
Process common cause failure considerations such as corrosion, plugging, coating, etc.	
Regulatory requirements impacting the SIS	
Other	
Detailed Requirements for Each SIF	
ID number of SIF	
Required SIL for each SIF	
Expected demand rate	
Test interval	
The definition of the safe state of the process, for each of the identified events	
The process inputs and their trip points	
The normal operating range of the process variables and their operating limits	
The process outputs and their actions	
The functional relationship between process inputs and outputs, including logic, math functions, and any required permissives	
Selection of de-energized to trip or energized to trip	
Consideration for manual shutdown	
Action(s) to be taken on loss of energy source(s) to the SIS	
Response time requirement for the SIF to bring the process to a safe state	
Response action to diagnostics and any revealed fault	
Human-machine interface (HMI) requirements	
Reset function(s)	
Requirements for diagnostics to achieve the required SIL	
Requirements for maintenance and testing to achieve the required SIL	
Reliability requirements if spurious trips may be hazardous	
Failure mode of each control valve	
Failure mode of all sensors and transmitters	
Other	

One simple but effective documentation tool is a cause and effect diagram. (See Table 5.2). This diagram shows the input/output relationship for the safety system, trip set points, SIL for each function, and any special design requirements.

Table 5-2: Cause and Effect Diagram

Tag #	Description	SIL	Instrument Range	Trip Value	Units	Open Valve FV-1004	Open Valve XV-1005	Stop Pump P-1007	Close Valve PV-1006	Notes
FALL-1000	Flow to main reactor R-100	2	0-200	100.0	GPM	X		X		1
PAHH-1002	Reactor internal pressure	2	0-800	600	PSI	X	X			1,2
XA-1003	Loss of control power					X	X	X	X	
XA-1004	Loss of instrument air					X	X	X	X	
HA-1005	Manual shutdown					X	X	X	X	

Notes

1. 2 second delay required before closing FV-1004
2. Reset located at valve PV-1006

The specifications should state *what* is to be achieved, not necessarily *how* to achieve it. This allows others the freedom to decide how best to accomplish the goals, and also allows for new knowledge and techniques.

5.2.9 Inadequate Final Review of Specification

The specifications should be reviewed upon completion and approved by all parties involved with the SIL determination. This will ensure understanding and agreement. Those responsible should not assume that everyone involved had the same understanding during the SIL review. This final review will give everyone an opportunity to reflect on the decisions and to clarify any misunderstandings.

5.2.10 Unauthorized Deviation from Specification

Once the specification is approved, no deviations should be made unless formal change procedures are followed. The system and specification may need to be updated if changes are made during the course of the project. (See Chapter 13, "Managing Changes to a System".) While it may be tempting to make certain changes in order to avoid cost or schedule hurdles, no changes should be made without a proper review and approval process.

5.3 ANSI/ISA-84.00.01-2004, Parts 1-3 (IEC 61511 Mod) Requirements

Subclause 10.3 of ANSI/ISA-84.00.01-2004 (IEC 61511 Mod) provides a summary of the SIS safety requirements. Below are the key requirements of the standard. Refer to the standard for additional details.

The requirements shall be sufficient to design the SIS and shall include the following:

- A description of all the SIFs necessary to achieve the required functional safety
- Requirements to identify and take account of common cause failures
- A definition of the safe state of the process for each identified SIF
- A definition of any individually safe process states which, when occurring concurrently, create a separate hazard (for example, overload of emergency storage, multiple relief to flare system)
- The assumed sources of demand and demand rate of each SIF
- Requirements for proof-test intervals
- Response time requirements for the SIF to bring the process to a safe state
- The SIL and mode of operation (demand/continuous) for each SIF
- A description of process measurements and their trip points
- A description of process output actions and the criteria for successful operation (e.g., requirements for tight shut-off valves)

- The functional relationship between process inputs and outputs, including logic, mathematical functions, and any required permissives
- Requirements for manual shutdown
- Requirements relating to energize or de-energize to trip
- Requirements for resetting the SIF after a shutdown
- Maximum allowable spurious trip rate
- Failure modes and desired response of the SIF
- Any specific requirements related to the procedures for starting up and restarting the SIF
- All interfaces between the SIS and any other system (including the BPCS and operators)
- A description of the modes of operation of the plant and identification of the SIFs required to operate within each mode
- The application software safety requirements (listed below)
- Requirements for overrides/inhibits/bypasses including how they will be cleared
- The specification of any action necessary to achieve or maintain a safe state in the event of fault(s) being detected in the SIF
- The mean time to repair which is feasible for the SIF
- Identification of the dangerous combinations of output states of the SIS that need to be avoided
- Identification of the extremes of all environmental conditions that are likely to be encountered by the SIS
- Identification to normal and abnormal modes for both the plant as a whole (e.g., plant startup) and individual plant operational procedures
- Definition of the requirements for any safety instrumented function necessary to survive a major accident event (e.g., the time required for a valve to remain operational in the event of a fire)

Subclause 12.2 of the standard provides requirements for the specification of the application software safety requirements. It's essential that the application software specifications be consistent with the safety requirements listed above.

- An application software safety requirements specification shall be developed.
- The input to the specification of the software safety requirements for each SIS subsystem shall include:
    - the specified safety requirements of the SIF
    - the requirements resulting from the SIS architecture, and
    - any requirements of safety planning
- The specification of the requirements for application software safety shall be sufficiently detailed to allow the design and implementation to achieve the required safety integrity and to allow an assessment of functional safety to be carried out.
- The application software developer shall review the information in the specification to ensure that the requirements are unambiguous, consistent and understandable.
- The specified requirements for software safety should be expressed and structured in such a way that they are clear, verifiable, testable, modifiable, and traceable.
- The application software safety requirements specification shall provide information allowing proper equipment selection.

5.4 Documenting the Specification Requirements

Since the specification will form the basis of the safety system design, all the required information should be included as a complete package. The following four items are key documents that should also be included in the SRS package.

1. The **process description**. This should include the following:
    - Piping and instrument drawings (P&IDs)
    - Description of the process operation
    - Process control description including control system design philosophy, type of controls, operator interface, alarm management, and historical data logging
    - Special safety regulations including corporate, local, state/provincial or federal requirements

- Reliability, quality, or environmental issues
- List of operational or maintenance issues

2. **Cause and effect diagram**. A cause and effect diagram can be used to document the functional and integrity requirements in a simple manner that is easily understood by all disciplines. See Table 5-2 for an example.
3. **Logic diagrams**. Logic diagrams can be used to supplement cause and effect diagrams for more complex and/or time based functions and sequences that may not easily be described using a cause and effect diagram. Logic diagrams should be produced in conformance with ISA-5.2-1976 (R1992).
4. **Process data sheets**. Process data sheets ensure that the information required for completing instrument specification sheets is well documented.

Summary

The safety requirement specifications (SRS) consists of two parts: the functional requirement specifications and the integrity requirement specifications. It includes both hardware and software requirements. The functional requirements describe the system inputs, outputs, and logic. The integrity requirements describe the performance of each safety function.

The English Health and Safety Executive examined 34 accidents that were the direct result of control and safety system failures in a variety of different industries. Their findings showed that 44% of accidents were due to incorrect and/or incomplete specifications (both the functional specifications as well as the integrity specifications).

The importance of ensuring that the system specifications are correct cannot be overemphasized.

References

1. Kletz, Trevor A. *Computer Control and Human Error.* Gulf Publishing Co., 1995.
2. ANSI/ISA-84.00.01-2004, Parts 1-3 (IEC 61511-1 to 3 Mod). *Functional Safety: Safety Instrumented Systems for the Process Industry Sector.*
3. Bird, Frank E., George L. Germain, and F.E. Bird, Jr. *Practical Loss Control Leadership*. International Loss Control Institute, 1996.

4. *Programmable Electronic Systems in Safety Related Applications, Part 1 - An Introductory Guide.*, U.K. Health & Safety Executive, 1987.

6

DETERMINING THE SAFETY INTEGRITY LEVEL (SIL)

Chapter Highlights

"The man who insists upon seeing with perfect clearness before deciding, never decides."

— H. F. Amiel

6.1 Introduction

Today's safety system standards are performance based, not prescriptive. They do not mandate a technology, level of redundancy, test interval, or system logic. Essentially they state "the greater the level of risk, the better the safety systems needed to control it." There are a variety of methods of evaluating risk. There are also a variety of methods of equating risk to the performance required from a safety system. One term used to describe safety system performance is safety integrity level (SIL).

Many industries have the need to evaluate and rank risk. Management decisions may then be made regarding various design options. For example, how remote, if at all, should a nuclear facility be to a large population zone? What level of redundancy is appropriate for a military aircraft weapons control system? What strength should jet engine turbine blades be for protection against flying birds? How long should a warranty period be based on known failure rate data? Ideally, decisions such as these would be made based upon mathematical analysis. Realistically, quantification of *all* factors is extremely difficult, if not impossible, and subjective judgment and experience may still be considered.

Military organizations were some of the first groups to face such problems. For example, when someone has to press the button that might start or stop World War III, one would like to think that the probability of the electronic circuitry working properly would be rated as something other than "high." The U.S. military developed a standard for categorizing risk: MIL-STD 882 "Standard Practice for System Safety" which has been adapted by other organizations and industries in a variety of formats.

Different groups and countries have come up with a variety of methods of equating risk to safety system performance. Some are qualitative, some more quantitative.

It's important to clarify what SIL is and what it is not. SIL is a measure of safety system performance. A system consists of a sensor, logic box, and final element(s). Therefore it is incorrect to refer to the SIL of an individual component of a system (e.g., a logic solver in isolation). A chain is only as strong as its weakest link. A logic solver could be rated for use in SIL 3, but if connected with non-redundant field devices with infrequent testing, the overall system may only meet SIL 1. There has been a push in industry for suppliers to use terms such as "SIL claim limit" and "SIL capability" in order to distinguish between component and system performance. It would be more appropriate to refer to the Probability of Failure on Demand (PFD) of an individual component. Unfortunately, PFD numbers are usually very small and referred to using scientific notation, which is difficult for some to relate to. Also, SIL is not a measure of process risk. It would be incorrect to say, "We've got a SIL 3 process."

6.2 Who's Responsible?

Determining SILs is mentioned in the various safety system standards. Many therefore assume the task falls upon the responsibility of the instrument or control system engineer. This is *not* the case. Evaluating the process risk and determining the appropriate safety integrity level is a responsibility of a multi-disciplined *team*, not any one individual. A control system engineer may (and should) be involved, but the review process will also require other specialists, such as those typically involved in a HAZard and OPerability study (HAZOP). Some organizations prefer to determine the SIL during the HAZOP. Others believe it's not necessary (or even desirable) to involve the entire HAZOP team and that the SIL determination should be performed as a separate study after the HAZOP. As a minimum, representatives from the following departments should participate in a SIL determination: process, mechanical design, safety, operations, and control systems.

6.3 Which Technique?

When it comes to hazard and risk analysis and determining safety integrity levels, there are no answers that could be categorized as either right or wrong. There are many ways of evaluating process risk, none more correct than another. Various industry documents describe multiple qualitative and quantitative methods for evaluating risk and determining SIS performance. [Ref. 2, 3, and 4] The methods were developed at different times by different countries. All methods are valid. Unfortunately, they may all also result in different answers.

Qualitative techniques are easier and faster to apply, but experience has shown they often provide conservative answers (i.e., high SIL requirements). This results in over-designing the safety functions and potentially unnecessary expenditures.

More quantitative techniques require greater effort in their development and usage, but experience has shown them to often provide lower SIL requirements. The difference in cost between a single SIL 2 versus a SIL 1 function can be tens of thousands of dollars. The difference between SIL 2 and 3 is even greater. Therefore, there may be significant cost savings associated with utilizing the more detailed quantitative techniques.

In general, if the use of qualitative techniques indicates no greater than SIL 1 requirements, continue their use. However, if the qualitative techniques indicate a significant percentage of SIL 2 or SIL 3 requirements, seriously consider the use of more quantitative techniques such as Layer Of Protection Analysis (LOPA). The overall cost savings that will result, especially over multiple projects, will easily justify their development and usage.

6.4 Common Issues

No matter what risk analysis method is chosen, several factors are common to all. For example, all methods involve evaluating the two components of risk—probability and severity—usually by categorizing each into different levels.

There are different hazardous events associated with each process unit and each event has its associated risk. Take for example a vessel where one is measuring pressure, temperature, level, and flow. The pressure measurement is probably meant to detect and prevent an over-pressure condition and explosion. This event would have a corresponding level of risk (i.e., probability and severity). Low flow might only result in a pump burning up. This would have a completely different probability and severity rating and, therefore, a different SIL requirement. High temperature might result in an off-spec product. This would also have a completely different probability and severity rating, and a different SIL requirement. What this means is one should *not* try and determine the SIL for an *entire* process unit or piece of equipment, but rather determine the SIL for *each safety function*.

6.5 Evaluating Risk

Risks are everywhere. Working in a chemical plant obviously involves risk, but then so does taking a bath. One involves more risk than the other, but there is a measure of risk in everything we do.

While zero injuries may be a goal for many organizations, it is important to realize that there is no such thing as zero risk. One is at risk merely sitting at home watching television (e.g., heart attack, flood, tornado, earthquake, etc.). So how safe should a process plant be? Should the risk of working at a chemical plant be equal to that of staying at home, or driving one's car, or flying in an airplane, or skydiving? All things being equal, the safer a plant is, the more expensive it will be. There has to be an economic consideration at some point. If the safety goal precludes the process from even being started, something obviously needs to be changed.

6.5.1 Hazard

The American Institute of Chemical Engineers (AIChE) definition of a hazard is "an inherent physical or chemical characteristic that has the potential for causing harm to people, property, or the environment. It is the combination of a hazardous material, an operating environment, and certain unplanned events that could result in an accident."

Hazards are always present. For example, gasoline is a combustible liquid. As long as there is no ignition source, gasoline may be considered relatively benign. Our goal is to minimize or eliminate hazardous events or accidents. We therefore don't store gasoline near ignition sources.

6.5.2 Risk

Risk is usually defined as the combination of the severity and probability of a hazardous event. In other words, how often can it happen, and how bad is it when it does. Risk can be evaluated qualitatively or quantitatively.

While many of the safety system standards would appear to focus on personnel risk, people are not the only thing at risk in process plants. For example, unmanned offshore platforms have a considerable amount of risk, even though there may be no personnel on the platform most of the time. Factors at risk are:

- **Personnel**. Both plant personnel and the nearby public.
- **Lost production downtime**. The loss associated with not being able to sell a product. This can often be the largest economic factor.
- **Capital equipment**. The costs associated with replacing damaged equipment.
- **Environment**. Polluting food supplies used by multiple countries or requiring the evacuation of an area is obviously something to be avoided.
- **Litigation costs**. Legal costs associated with injuries, deaths, and environmental losses can be significant.
- **Company image**. Bad press can put a company out of business.

Some of these items (e.g., lost production, capital equipment) may be quantified, some others (e.g., company image) may be more difficult to quantify.

Safety system performance levels can be assigned based on any of the risks listed above. For example, there are organizations that have safety integrity level (SIL) based on personnel risk, commercial integrity level (CIL) based on economic issues, and environmental integrity level (EIL) based on environmental factors.

6.5.3 Fatality Rates

The most obvious and serious risk to personnel is a fatality. How safe should personnel in or nearby a process plant be? Should working in a process plant be as safe as driving a car? In the U.S. there are approximately 45,000 fatalities every year due to automobile accidents. Would having a similar risk in the process industry mean the industry could kill 45,000 people every year? Obviously not! For one, not as many people work in process plants as drive automobiles. The issue is not total number of fatalities, but fatality rates.

There are two common methods of expressing fatality rates. One is the fatal accident rate, or FAR. It is the number of deaths per million person hours of exposure. Another is the probability per unit of time. There are a number of sources that list fatality rates for various activities, from different industries, means of transportation, recreational activities, including voluntary as well as involuntary risks. Table 6-1 lists fatality rates for various activities in the U.K. [Ref. 1] Note that probabilities for two different activities may be the same, yet the FAR may be different, due to the difference in exposure hours. Also note that the numbers come from different sources and all of the underlying information may not be available, so there is not always the means of directly relating probabilities to rates, hence some of the fields in Table 6-1 are blank.

Table 6-1: Fatal Accident Rates (in the United Kingdom)

Activity	Probability (per year)	FAR
Travel		
Air	2×10^{-6}	
Train	2×10^{-6}	3 - 5
Bus	2×10^{-4}	4
Car	2×10^{-4}	50-60
Motorcycle	2×10^{-2}	500 - 1,000
Occupation		
Chemical Industry	5×10^{-5}	4
Manufacturing		8
Shipping	9×10^{-4}	8
Coal Mining	2×10^{-4}	10
Agriculture		10
Boxing		20,000
Voluntary		
The Pill	2×10^{-5}	
Rock Climbing	1.4×10^{-4}	4,000
Smoking	5×10^{-3}	
Involuntary		
Meteorite	6×10^{-11}	
Falling Aircraft	2×10^{-8}	
Firearms	2×10^{-6}	
Cancer	1×10^{-6}	
Fire	2.5×10^{-5}	
Falls	2×10^{-5}	
Staying at Home	1×10^{-4}	1 – 4

6.5.4 Risks Inherent in Modern Society

Just how safe is "safe"? Should working in a plant have the same level of risk as skydiving (which kills about 40 people per year in the U.S.)? Should working in a plant be as safe as driving your car? Or should it be as safe as flying in a plane, which is safer than driving a car by two orders of magnitude?

While the term FAR may be simple to understand and may represent a useful yardstick, many companies, especially in the U.S., are unwilling to put such targets in writing. Imagine walking into company XYZ's plush world headquarters office and on the wall in the reception area is a sign that reads, "We at XYZ consider it tolerable to kill 4 people per 100 million man hours." The lawyers would have a field day! However, as we shall see, some organizations *have* established such quantified risk targets.

People's perception of risk varies depending on their understanding or familiarity with the risk. For example, most people are familiar with driving. The perceived level of risk is relatively low to most people, even though approximately 45,000 people die every year in the U.S. alone due to traffic accidents. If a new chemical facility is being proposed nearby a residential area, the level of understanding of the residents regarding the chemical process will probably be low, and their discomfort, or perceived level of risk will no doubt be high, even if the process may have a very good historical safety record.

Perception of risk will also vary in proportion to the number of possible deaths associated with a particular event. For example, of the 45,000 traffic fatalities every year, the deaths usually occur one, or a few, at a time. Even with this surprisingly high number of deaths, there's little (if any) public outcry that something be done to lower this figure. Yet when there's an accident with a school bus involving injuries to many children there typically is an outcry. The same could be said about relatively high-risk sports such as skydiving, hang gliding, and ultralight aircraft. Although these sports involve relatively high risk, it's rare that one hears of multiple fatalities. The people involved also made their own conscious choice to partake in the activity and outsiders are generally not exposed to the risks. Accidents involving the chemical industry, however, frequently do involve multiple fatalities. Bhopal was the worst to date with over 3,000 deaths and 200,000 injuries. The overall risk associated with working in a chemical plant can be shown to be less than the risk of driving (at least in the U.S.), yet the public's *perception* of the risks of the two activities is typically reversed.

6.5.5 Voluntary vs. Involuntary Risk

There's a difference between voluntary and involuntary risks. Examples of voluntary risks would be driving a car, smoking cigarettes, and so on. Examples of involuntary risks would be having a chemical plant built near your home after you've lived there a number of years or secondary smoke from other peoples' cigarettes.

People can perceive similar risks differently. For example, the Jones own a house in the country. One day, company XYZ builds a toxic chemical plant nearby. After the plant is built, the Smiths buy a house next to the Jones. Both households face the same risk, but they'll probably each have a different perception of it. To the first couple (the Jones) who lived there before the plant was built, the risk is involuntary (although they obviously could move). To the second couple (the Smiths), who bought their home after the plant was built, it's voluntary.

People are usually willing to face higher voluntary risks than involuntary ones. For example, when one of the authors was younger he was willing to accept both the risks of riding a motorcycle in a large city and of skydiving. The risks were voluntary and he considered himself to be the only one at risk at the time. (One could argue the finer points of that.) As a married father, he no longer wishes to accept those risks (never mind the fact that he can no longer afford them).

Another factor involved in the perception of risk is control. For example, the wife of one of the authors does not like flying. (In fact, flying is the number two fear among Americans. Public speaking is number one!) Her stated reason for discomfort is that she doesn't feel "in control." When you're sitting behind the wheel in your car at a stop sign and a drunk driver plows into your car, you weren't in control then either. After all, no one goes out planning to have an accident.

6.5.6 Tolerable Levels of Risk

The concept of acceptable or tolerable levels of risk is not solely a technical issue; it involves philosophical, moral and legal matters. Deciding how safe is safe enough can't be answered by algebraic equations or probabilistic evaluations. Alvin Weinberg has termed these "transscientific questions," for they transcend science. [Ref. 7]

One issue that presents difficulties is trying to statistically estimate extremely unlikely events. Estimating very rare events, such as a severe chemical accident, cannot have the same validity as estimates for which abundant statistics are available. Because the desired probabilities can be so small (e.g., 10-6 per plant per year), there is no practical means of determining the rate directly. One can't build ten thousand plants and operate them for one hundred years in order to tabulate their operating histories. Putting it a simpler way, measuring something that doesn't happen very often is difficult.

6.5.7 Tolerable Risk in the Process Industries

It's common to view personal risk in a subjective, intuitive manner. Many people won't consider driving a motorcycle, no matter how wonderful their biker friends say it may be. The author's wife who doesn't like flying believes it's all right for her family to drive to the airport in the same car but not to fly in the same airplane (even though she understands that flying is two orders of magnitude safer). Logic doesn't always apply when evaluating relative risk.

We should not, however, have the same subjective attitude about risks in the process industry. Usually, the people making the risk decisions (e.g., engineers) are not the ones who will be facing the eventual risk (e.g., workers or nearby residents). Although none of the more famous accidents in the process industry would ever be considered "acceptable," the companies involved did not go out of business the next day. Therefore, the losses must have been considered "tolerable." How many accidents might the industry be willing to consider tolerable? How many accidents must take place before there's public and political outcry?

There are approximately 2,300 petrochemical plants in the United States alone. If an average of one were to have a catastrophic event involving over 20 fatalities every year (which represents an individual plant risk of 1/2,300 per year), how long would it be before there was a public outcry and the government stepped in? What if such an accident only happened once every ten years (1/23,000 per year)? There is no such thing as zero risk, but it's very difficult to decide what particular level should be considered "tolerable."

The relevant statistics can be rather confusing. An individual risk of 1/2,300 per year means that out of 2,300 plants, on average, one might go "boom" every year. It's important to realize, however, that you can't predict *which* plant and you can't predict *when* one will go "boom." But since people don't build 2,300 plants all at once, or live next to 2,300 plants, they want to know the risk of the one plant they're associated with. The risk for an individual plant remains the same, 1/2,300 per year. However, some are just not comfortable with such a number. Some twist things around a bit and say the risk of an accident is "once every 2,300 years." This causes even more confusion. Some then assume it will be 2,300 years before there's an accident and, therefore, they have nothing to worry about. Nothing could be further from the truth. For example, approximately 1 in 4,000 people in the U.S. die in a car crash every year. If you go to a sports event with 4,000 people present, one can make the assumption that someone will die in an automobile accident within the next 365 days. However, you can't predict *which* person and you can't predict which *day*. The error associated with inverting the number and stating that you'll live 4,000 years before you die in a car crash should now be obvious.

Deciding what level of risk is tolerable could be compared to choosing your weapon in Russian roulette—how many barrels do you want in your gun? Would you choose an automatic pistol that always had a round in the chamber? (I hope not, although there have been such Darwin award winners!) Or would you choose a revolver with six chambers? What if you could choose a gun with fifty barrels or one with five thousand barrels? Obviously, the more barrels there are, the lower the risk of hitting the bullet when you pull the trigger. Either way, you have to play the game. You do, however, at least get to choose your weapon. There is no such thing as zero risk.

One in 2,300 per year, or the incorrect reciprocal of 2,300 years between accidents, may initially sound so remote as to be of no concern to most people. A more intuitive answer might initially be fifty years. (The reasoning being that someone's working life is fifty years, and they don't want anything to happen during *their* life.) But how long will the plant actually be around? Let's say 25 years. So take a gun with fifty barrels and pull the trigger once a year. What's the likelihood of hitting the bullet? 50%. (Although based on another set of assumptions and simplifications, the answer is 40%. Even the statistics involved are not inherently obvious!)

Would you want to have your name on the drawings of a plant if you knew there was a 50% chance of a catastrophic accident happening during the life of the plant? Probably not.

What if instead of 50 years, you choose 500 years? Then the risk would now be 25/500, or a 5% chance. Should that be tolerable? What about 5,000 years? Now it becomes 0.5%. There is no such thing as zero risk, but just how low must one go for the risk to be considered tolerable? That's the proverbial $64,000 question to which there is no scientific answer.

U.S. industry cannot afford to blow up an entire plant, involving dozen of fatalities, once every year (1/2,300). The negative press and public outcry would be ruinous. After all, OSHA stepped in and produced 29 CFR

1910.119 after several major Gulf Coast accidents happened over approximately a ten-year period. A serious accident once every ten years *might* be viewed as tolerable (1/23,000). Risk targets in the range of 1/10,000 have been documented. In fact, in The Netherlands, the government even publishes what it considers to be tolerable fatality rates. Switzerland and Singapore have done the same. Fatality rates are typically in the range of 1/1,000,000 per year. [Ref. 8]

6.6 Safety Integrity Levels

The greater the level of process risk, the better the safety system needed to control it. Safety integrity level is a measure of safety system performance. It is *not* a direct measure of process risk.

Standards and guidelines have categorized four overall safety integrity levels. Note that previous U.S. process industry documents only went up to SIL 3. The basis for the performance goals listed in Tables 6-2 and 6-3 have been a subject of controversy. In other words, how were these tables "calibrated"? Perhaps the numbers should be shifted up a row, or perhaps down a row. Why start at 90% and not 80%? Suffice it to say that the tables were accepted by different industry groups and appear in all the standards.

Table 6-2: Safety Integrity Levels and Required Safety System Performance for Low Demand Mode Systems

Safety Integrity Level (SIL)	Probability of Failure on Demand (PFD)	Safety Availability (1-PFD)	Risk Reduction Factor (1/PFD)
4	.0001 - .00001	99.99 – 99.999%	10,000 – 100,000
3	.001 - .0001	99.9 - 99.99%	1,000 - 10,000
2	.01 - .001	99 - 99.9%	100 - 1,000
1	.1 - .01	90 - 99%	10 - 100

Table 6-3: Safety Integrity Levels and Required Safety System Performance for Continuous Mode Systems

Safety Integrity Level (SIL)	Dangerous Failure per Hour
4	$\geq 10^{-9} < 10^{-8}$
3	$\geq 10^{-8} < 10^{-7}$
2	$\geq 10^{-7} < 10^{-6}$
1	$\geq 10^{-6} < 10^{-5}$
Notes: 1. Field devices are *included* in the above performance requirements. 2. Refer to Section 3.3.2 of this text for a discussion of low vs. continuous mode systems	

Different people use different terms to describe system performance. A very common term, especially among control system vendors, is the generic term 'availability'. The ISA SP84 committee purposely used a different term since the safety standards are referring to a *completely* different type of performance. This subject is discussed in more detail in Chapter 8. Safety availability can still be a confusing term as most figures are so close to 100%. For example, do you consider there to be a significant difference between 99% and 99.99%? The difference could be viewed as being less than 1%.

The compliment of safety availability is referred to as the probability of failure on demand (PFD). Now the numbers are so small that one must use scientific notation. How many plant managers will understand when their control system engineer tells them that their safety instrumented function has a PFD of 4.6×10^{-3}?

The reciprocal of PFD is called the risk reduction factor (RRF). The benefit to this term is that the difference between numbers is easier to see. For example, the difference between a risk reduction factor of 100 and 10,000 is obviously two orders of magnitude. Actually, the difference between a safety availability of 99% and 99.99% is also two orders of magnitude, although most don't realize it initially.

The following sections describe the various methods of determining the required safety integrity level based on process risk issues. The methods are described in both the IEC 61508 and 61511 standards and other texts. Please keep in mind that no SIL determination methodology is better or more correct than another. They were simply developed by different groups around the world at different times. Similarly, a Peugeot is probably no better than a Ford. Both are simply cars developed by different countries. No doubt national pride will come into play when qualifying each, but both serve their purposes equally well.

6.7 SIL Determination Method #1: As Low As Reasonably Practical (ALARP)

The English developed the theory of ALARP which is considered a *legal* concept. While not a method for determining safety integrity levels directly, it is a concept that can still be applied in this area. The idea hinges on three overall levels of risk and the economics associated with lowering the risk. The three overall levels of risk are defined as "unacceptable", "tolerable", and "broadly acceptable" as shown in Figure 6-1.

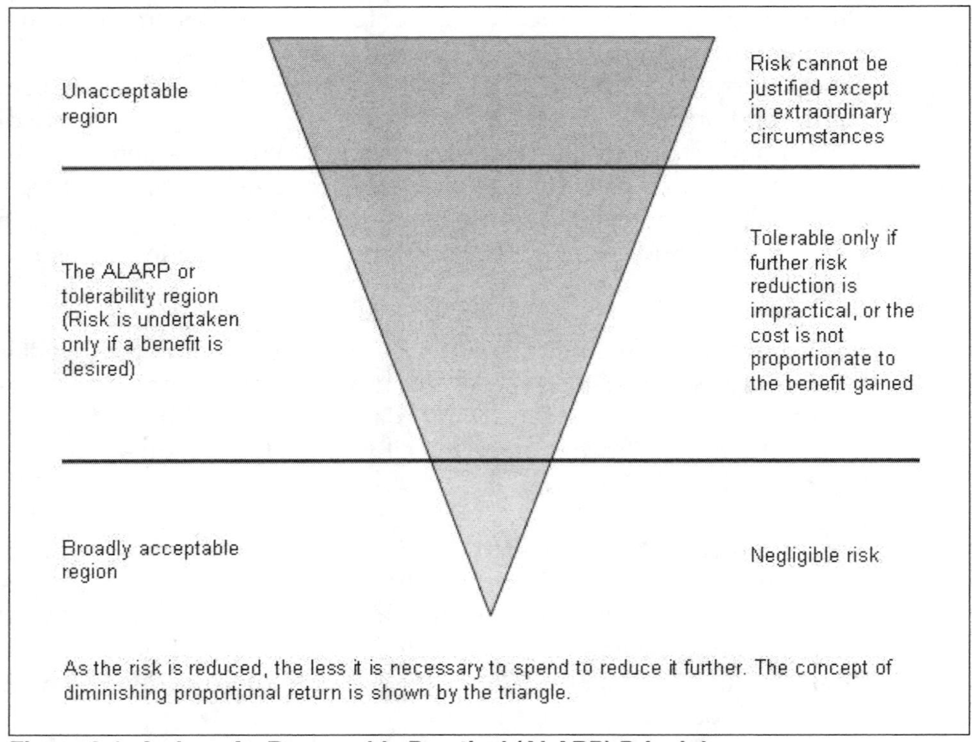

Figure 6-1: As Low As Reasonably Practical (ALARP) Principle

"Unacceptable" risks are viewed as so high as to be intolerable. For example, if a risk analysis showed the probability of destroying a large, manned, deepwater offshore platform at once per year, it's doubtful the design would proceed. Something must be done to lower the risk to a more tolerable level, regardless of cost.

"Broadly acceptable" risks are ones that are so low as to not be of major concern. For example, it is possible for a refinery to be hit by a meteor or terrorist airplane attack. However, building a missile defense system around every process plant would hardly be considered warranted or economically feasible.

In between these two extremes are risks that are considered "tolerable" if the positive benefit is seen to outweigh the potential negative impact. Driving a car is an excellent example. The fact that approximately 45,000 people in the U.S. die in auto accidents every year does little to deter the rest of us from driving.

While such decisions are primarily economic, other factors do influence company judgments. For example, no nuclear power plants have been built in the U.S. for over two decades. Similarly, few grassroots chemical plants are being built in the U.S., no matter how safe they may be shown to be, or no matter how positive the economic factors for a community may be. The combined power of Not In My Back Yard (NIMBYs) has influenced many companies to build their facilities overseas where public, political, legal, and economic conditions are more favorable.

6.8 SIL Determination Method #2: Risk Matrix

The risk matrix found in many standards, recommended practices, and internal company procedures has many similarities to a U.S. military standard (MIL STD 882). Both categorize frequency and severity of a hazardous event using multiple qualitative levels. The risk matrix common in the process industry, however, often adds a third axis to account for additional safety layers.

6.8.1 Evaluating the Frequency

The frequency or probability of an event may be ranked from low to high, improbable to frequent, or whatever term may be appropriate. This may be for a single item, or for a group of items, a single process unit, or an entire plant. Levels may be ranked qualitatively or quantitatively as shown in Table 6-4. If quantitative values are chosen, it's suggested that they differ by at least an order of magnitude, since the safety integrity levels differ by single orders of magnitude.

Table 6-4: Risk Frequency (Example Only)

Level	Descriptive Word	Qualitative Frequency	Quantitative Frequency
5	Frequent	A failure that can reasonably be expected to occur more than once within the expected lifetime of the plant	Freq > 1/10 per year
4	Probable	A failure that can reasonably be expected to occur within the expected lifetime of the plant	1/100 < Freq < 1/10 per year
3	Occasional	A failure with a low probability of occurring within the expected life of the plant.	1/1,000 < Freq < 1/100 per year
2	Remote	A series of failures, with a low probability of occurring within the expected life of the plant	1/10,000 < Freq < 1/1,000 per year
1	Improbable	A series of failures, with a very low probability of occurring within the expected life of the plant	Freq < 1/10,000 per year

6.8.2 Evaluating the Severity

Severity may also be categorized according to the different factors at risk; people, capital equipment, production, etc. The numbers shown in Tables 6-4 and 6-5 are *examples only* and are *not* intended to represent, or even imply, any kind of recommendation. The monetary values shown in Table 6-5 are also examples only and need to be "calibrated" for each organization. For example, what may be a catastrophic financial loss to one organization may be a relatively trivial loss to another.

Table 6-5: Risk Severity (Example Only)

Level	Descriptive Word	Potential Severity/Consequences		
		Personnel	Environment	Production/ Equipment
V	Catastrophic	Multiple fatalities	Detrimental offsite release	Loss > $1.5M
IV	Severe	Single fatality	Non-detrimental offsite release	Loss between $1.5M and $500K
III	Serious	Lost time accident	Release onsite - not immediately contained	Loss between $500K and $100K
II	Minor	Medical treatment	Release onsite - immediately contained	Loss between $100K and $2,500
I	Negligible	First aid treatment	No release	Loss < $2,500

6.8.3 Evaluating the Overall Risk

The two sets of numbers may then be combined into an X-Y plot as shown in Table 6-6. The lower left corner represents low risk (i.e., low probability and low frequency), the upper right corner represents high risk (i.e., high probability and high frequency). Note that defining the borders between categories is rather subjective. One may also use more than three overall categories.

These risk levels are similar to the ALARP concept. If high risk is identified, process changes may be called for, whether the facility is new or existing. If low risk is identified, no changes may be warranted. If moderate risk is identified, additional safety layers or procedures may be necessary. Note that these are merely examples.

Table 6-6: Overall Risk (Example Only)

	Frequency				
Severity	**1**	**2**	**3**	**4**	**5**
V	1-V	2-V	3-V	4-V	5-V
IV	1-IV	2-IV	3-IV	4-IV	5-IV
III	1-III	2-III	3-III	4-III	5-III
II	1-II	2-II	3-II	4-II	5-II
I	1-I	2-I	3-I	4-I	5-I

High Risk

Low Risk

Medium Risk

The two dimensional matrix can be used to evaluate overall risk. It should *not*, however, be used to determine safety integrity levels as there are additional process design issues that should be accounted for, such as additional safety layers, as covered in the next section.

6.8.4 Evaluating the Effectiveness of Additional Layers

A third axis appears in a variety of standards and guidelines which does not appear in the military standard (see Figure 6-2). This third axis is designed to take into account the additional safety layers commonly found in the process industries. Note that the Z axis is labeled "quantity and/or effectiveness of additional layers." This refers to layers *out*side of the safety instrumented system shown in the "onion diagram" (see Figure 4-2). In other words, if the safety system under consideration were to fail, might any other layer(s) be able to prevent or mitigate the hazardous event? Note that the basic process control system, alarms, and operators are already assumed to be in place and should *not* be accounted for in the Z axis. If, for example, a redundant process control system were being considered, it might be possible to consider a lower frequency/probability on the X axis. This would essentially have the same overall effect of lowering the required safety integrity level of the safety system.

The numbers shown in the squares in Figure 6-2 represent safety integrity levels and are merely examples. Which number is placed in which box is *not* consistent across industry and needs to be settled on within each organization. For example, previous U.S. standards did not recognize SIL 4. Just because SIL 4 is now defined in more recent standards does not necessarily mean that organizations need to modify previous graphs, especially if they felt that they did not even have any SIL 3 requirements.

One difficulty associated with Figure 6-2 (as well as Table 6-4) is that frequencies should be chosen assuming the safety layers are *not* in place. If the hazardous event under consideration were an overpressure of a vessel resulting in an explosion, the historical records of vessel explosions would already take into account safety layers such as safety systems, pressure relief valves, rupture discs, etc. Estimating what hazardous event frequencies *might* be with*out* these layers proves to be a difficult exercise.

As an example, suppose the hazardous event under consideration were an overpressure condition resulting in an explosion of a vessel. If the area were normally manned, there might be multiple fatalities. This might result in the highest severity rating. Historical records of vessel explosions might indicate a very low frequency, although the records already account for various safety layers. Again, this is merely a hypothetical example. If there were no additional safety layers beyond the safety instrumented system, the Z axis cannot be utilized. This would result in a SIL 3 requirement as shown in Figure 6-2. However, if there are additional layers, such as a relief valve, one could utilize the Z axis and lower the design requirements of the safety system under study from SIL 3 to SIL 2. This would be accounting for the relief valve as a single layer. Note, however, that the Z axis can also be used to account for the *effectiveness* of additional layers. A quantitative analysis of a relief valve might show it to have an effectiveness equal to two layers. (This concept will be covered in more detail in the layer of protection analysis [LOPA] section.) This means the required safety integrity level might be lowered from SIL 3 down to SIL 1.

One concern using this safety matrix method has to do with the qualitative judgments involved. Every company can take the liberty of defining the levels and rankings a bit differently. Occupational Safety and Health Administration (OSHA) representatives have admitted that they are not process experts. They are not about to tell a user that they should have selected SIL 3 instead of SIL 2 for a particular function. They just want to see the paperwork trail your company has left behind and the decision process of how you got there. The fact that different locations within the same company can define all of this differently does not really matter (at least to the standards bodies), even though it is naturally cause for concern. The industry simply has not reached, and may never reach, consensus on all of the subjective decisions and ranking involved.

There's a simple yet rather crass thought to keep in the back of your mind during this whole process, "How will we defend this decision in a court of law?" As scary as that thought may seem, it sums it up rather well. Think about it. If there were an accident and people were injured, the case would most likely go to court. Someone somewhere will need to be able to defend their design decisions. If asked why a certain decision was made and the response is, "Well, we weren't sure what to do, but that's what our vendor recommended." the court's response may naturally be less than forgiving.

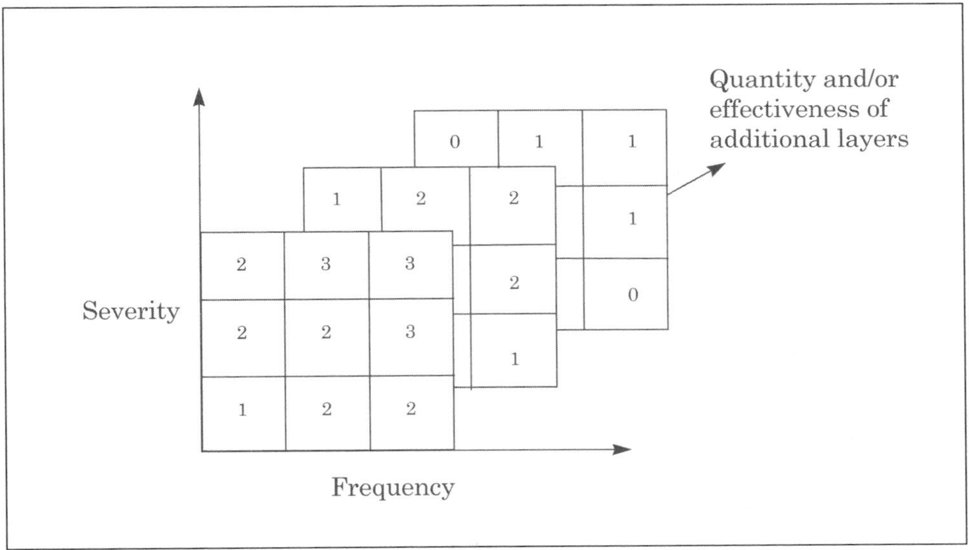

Figure 6-2: 3-Dimensional Risk Matrix for Selecting SIL

6.9 Method #2: Risk Graph

The Germans and Norwegians developed different, yet similar, techniques for determining safety integrity levels (see Figure 6-3). The wording used in the figure is purposely vague and open to interpretation. Each organization needs to define the boundaries/classifications more precisely. The standards are simply trying to show the methodology, not a working example that might be misused.

Figure 6-3 was intended for the risk to personnel. Similar graphs can be developed for risks associated with equipment, production losses, environmental impact, etc.

One starts at the left end of the chart. First, what are the consequences to the people involved for the hazardous event being considered? What if there are four fatalities? Would that be considered "several deaths" or "many deaths"? Again, each organization is expected to define the boundaries more precisely.

Next, what's the frequency and exposure of the personnel to the particular hazardous event? Keep in mind that even though a process might be con-

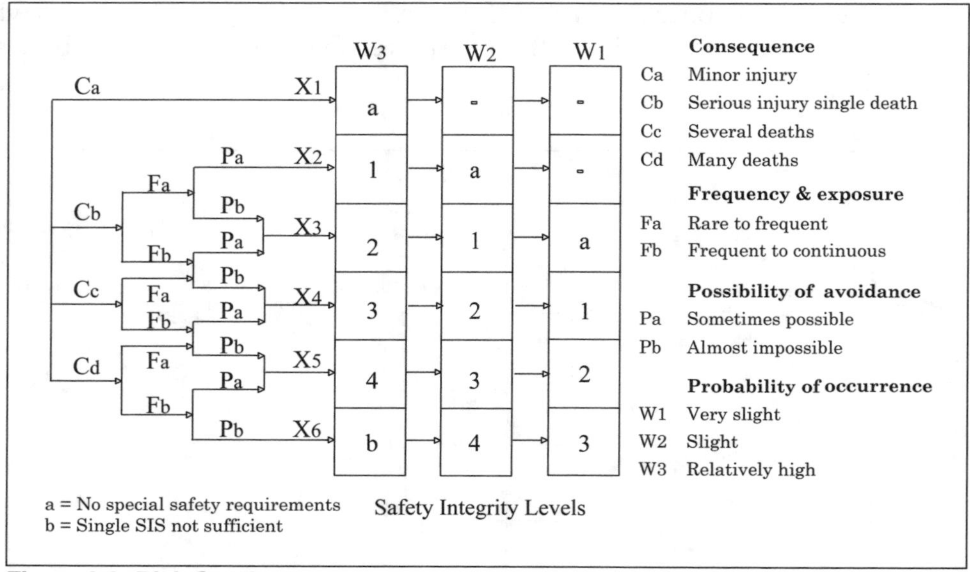

Figure 6-3: Risk Graph

tinuous, the personnel might only be in the particular area 10 minutes out of the day, so their *exposure* would *not* be continuous. One should also consider the increased manning that might occur during a hazardous event buildup. Frequent to continuous exposure may be rare. One obvious example might be living and working in a city with a nuclear power plant.

Next, one considers the possibility of avoiding the accident. There are multiple factors one may account for here. For example, what is the "process safety time," which is essentially defined as the time it takes for a hazardous event to occur if all control were lost? Is the process reaction time slow enough for the people to take action, or are the process dynamics so fast that operator intervention might not be possible? Do the operators have local indications to know what is going on in the process (so they can take corrective action)? Are personnel adequately trained to know what to do when an abnormal situation develops (so they may attempt to prevent it)? Are evacuation routes accessible in order to vacate the area in case the hazardous event were to occur? The differences between "sometimes possible" and "almost impossible" should be more clearly defined by each organization.

Finally, one considers the probability of the event actually occurring, again assuming the safety system were not in place. This primarily has to do with the additional safety layers and historical records of similar incidents. In other words, considering all of the other independent safety layers, what's the actual likelihood of this particular event happening?

The differences between "very slight," "slight," and "relatively high" should be more clearly defined by each organization.

One can see that all of this is rather subjective. One may also question the calibration of Figure 6-3. Who's to say this chart should also not be shifted up or down one row? No doubt the Europeans had some sort of rationale in its development.

6.10 Method #3: Layer of Protection Analysis (LOPA)

Some engineers are not comfortable with the purely qualitative methods outlined above for determining SIL. The repeatability of the methods has been questioned by some. For example, if you give the same problem to five different design groups and two groups come up with SIL 1, two groups with SIL 2, and one group with SIL 3, what might you conclude? Such differences would result in vastly different SIS designs. Are some of the groups "right" and others "wrong"? Actually, defining the boundaries and training the teams properly should make the overall determination process more consistent and repeatable.

Some are more comfortable with a more quantitative method. For example, if your company does not want the hazardous event to happen more often than once in a thousand years, and experience shows there to be a process demand on average once a year, then it's a relatively straightforward matter to determine the performance required of your safety layers.

Two AIChE CCPS textbooks describe LOPA. [Ref. 3 and 5] The onion diagram shown in Figure 4-2 illustrates how there are multiple layers of protection in any process plant. Each layer will have its own associated level of performance, or risk reduction. (Note that determining the actual performance of any one layer is the subject of Chapter 8.) If one knows a) the overall safety target (i.e., the tolerable risk level), b) what level of risk one is starting from (i.e., the initiating event frequency), c) how many safety layers are in place to prevent or mitigate the event, and d) the level of performance of each layer, one can calculate and determine if the overall target level of risk is being achieved. The concept is relatively easy to understand and is shown in Figure 4-3. However, coming up with numbers for all of the above is a bit of a challenge.

6.10.1 Tolerable Risk

The first step involves determining a tolerable level of risk. This topic was discussed earlier in Section 6.5. The legal system in the U.S. makes determining and documenting such numbers problematic; however, it has been done. Also, certain governments around the world have documented such a number for industry to use. Table 6-7 (from Ref. 5) lists examples of yearly tolerable fatality probabilities. Company names have been removed from this table. Similar tables can be developed for tolerable probabilities for different size losses due to the impact of environmental, lost production downtime, capital equipment, and other events.

Table 6-7: Sample Tolerable Risk Probabilities

	Max tolerable risk for workforce (all scenarios)	Negligible risk for workforce (all scenarios)	Max tolerable risk for public (all scenarios)	Negligible risk for public (all scenarios)
Company A	10^{-3}	10^{-6}	-	-
Company B	10^{-3}	10^{-6}	-	-
Company C	3.3×10^{-5}	-	10^{-4}	-
Company D	2.5×10^{-5} (per employee)	-	10^{-5}	10^{-7}
Typical (one scenario)	10^{-4}	10^{-6}	10^{-4}	10^{-6}

6.10.2 Initiating Event Frequencies

The next step involves determining initiating event frequencies or probabilities. These can either be external events (e.g., a lightning strike) or failure of one of the layers (e.g., a control system valve fails open leading to a hazardous event development). Numbers need to be determined for any event under consideration. Table 6-8 (from Ref. 5) as an example. Numbers may be determined based on historical records or failure rate data (which is one form of historical record).

Table 6-8: Sample Initiating Event Frequencies

Initiating Event	Frequency (per year)
Gasket / packing blowout	1×10^{-2}
Lightning strike	1×10^{-3}
BPCS loop failure	1×10^{-1}
Safety valve opens spuriously	1×10^{-2}
Regulator failure	1×10^{-1}
Procedure failure (per opportunity)	1×10^{-3}
Operator failure (per opportunity)	1×10^{-2}

Note that while operating and maintenance procedures may be considered by some to be a safety layer, it is recognized that errors will be made and procedures will be violated. Human issues such as these may in *some* cases be considered initiating events, in others they may be prevention layers.

6.10.3 Performance of Each Safety Layer

The next step is determining the level of performance of each safety layer. There are rules on what may be considered an independent protection layer (IPL). For example:

1. **Specificity**: An IPL is designed solely to prevent or to mitigate the consequences of one potentially hazardous event. Multiple causes may lead to the same hazardous event. Therefore, multiple event scenarios may initiate action of one IPL.
2. **Independence**: An IPL is independent of the other protection layers associated with the identified danger. The failure of one layer will not prevent another layer from working.
3. **Dependability**: It can be counted on to do what it was designed to do. Both random and systematic failures modes are addressed in the design.
4. **Auditability**: It is designed to facilitate regular validation of the protective functions. Proof testing and/or maintenance is necessary.

Table 6-9 (from Ref. 5) shows as an example of performance that may be assigned to different levels. The probabilities are primarily based on failure rates and test intervals. (Chapter 8 shows examples of how such probabilities may be calculated.)

Table 6-9: Sample PFDs for Different Independent Protection Layers

Passive Independent Protection Layers	**Probability of Failure on Demand**
Dike	1×10^{-2}
Fireproofing	1×10^{-2}
Blast wall / bunker	1×10^{-3}
Flame / detonation arrestors	1×10^{-2}
Active Independent Protection Layers	
Relief valve	1×10^{-2}
Rupture disk	1×10^{-2}
Basic Process Control System	1×10^{-1}

Considering human action as a safety layer can be difficult and controversial since human errors are well noted. In order to account for human as a safety layer, all of the following conditions must be met:

- their intended procedures must be documented,
- personnel must be trained on the specific procedures,
- personnel must have sufficient time to act (e.g., 15 minutes),
- the use of procedures must be audited

Even with the above conditions satisfied, a probability limit (cap) on human error of 1×10^{-1} is a reasonable. In other words, claiming that a person will respond correctly in a high stress emergency situation more often than 9 out of 10 times (even with the above four cases satisfied) may be considered unrealistic. The use of specialized techniques for evaluating human reliability would be required in order to make such a claim, and such techniques are rarely used in the process industries.

Table 6-9 is obviously just a partial listing of potential passive and active safety layers. A fire and gas system may be one of the safety layers and could have its performance quantified (which is the topic of Chapter 8). In many cases the performance of F&G systems will be limited to a PFD of approximately 1×10^{-1}. If the F&G system is to be considered a safety system and have an integrity level assigned (according to the procedures outlined in this chapter), then its performance (PFD) requirements may have to be considerably greater.

6.10.4 Example Using LOPA

The following example is a slight modification of one covered in Reference 5. Figure 6-4 shows the sample process. A vessel is used to store hexane, a combustible material. The level in the vessel is controlled by a level controller which operates a valve. If the vessel is overfilled, hexane will be released through a liquid vent and be contained within a dike. A hazard analysis was performed and it was determined that the level controller might fail, liquid might be released outside of the dike, an ignition source might ignite the hexane, and there might be a possible fatality. The organization wanted to determine if the existing facility would meet their corporate risk criteria, or if any changes were required (such as adding a stand-alone safety system), how extensive the changes would need to be.

The organization established a yearly tolerable risk limit for a fire of 1E-4 and 1E-5 for a fatality. The initiating event for this scenario would be a failure of the control system, which was estimated at 1E-1. The only existing safety layer would be the dike, which had an estimated PFD of 1E-2.

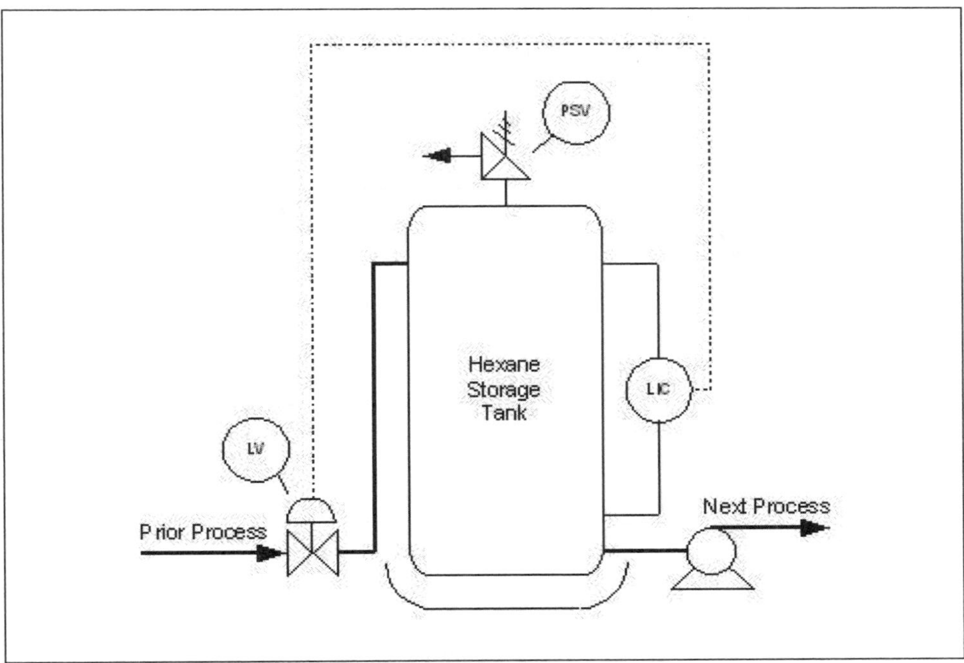

Figure 6-4: Sample Process for LOPA Example

Alarms and operator action were *not* accounted for because, in this instance, the control system was the initiating event, therefore no alarms would be generated. The organization took a conservative view that if material was released outside of the dike, the likelihood of it finding an ignition source would be 100%. However, the area was not always manned. The probability of someone being in the area was estimated at 50%. The probability of someone in the area actually being killed by a fire, as opposed to merely injured, was estimated at 50%.

Figure 6-5 shows an event tree for this scenario. The probability of a fire is represented by the combination of probabilities of the bottom three rows, which amounts to 1E-3 (0.1 x 0.01 x 1.0). The probability of a fatality is represented by the bottom row, which amounts to 2.5E-4 (0.1 x 0.01 x 1.0 x 0.5 x 0.5).

Knowing that the corporate risk target for a fire is 1E-4, we can see that the risk target is not being met by a factor of 10 (1E-3 / 1E-4). Knowing that the corporate risk target for a fatality is 1E-5, we can see that the risk target is not being met by a factor of 25 (2.5E-4 / 1E-5). Therefore, the existing design does *not* meet either corporate risk target and a change is warranted. One possible solution would be to install a separate high level shutdown function. Such a function would need to reduce the risk by at least a factor of 25 in order to meet the overall criteria. A risk reduction factor (RRF) of 25 falls within the SIL 1 range (10-100) as shown in Table 6-2. However, obviously not *any* SIL 1 system will do. The new system must meet a Risk Reduction Factor of at least 25. Figure 6-6 shows a sample worksheet for documenting this scenario.

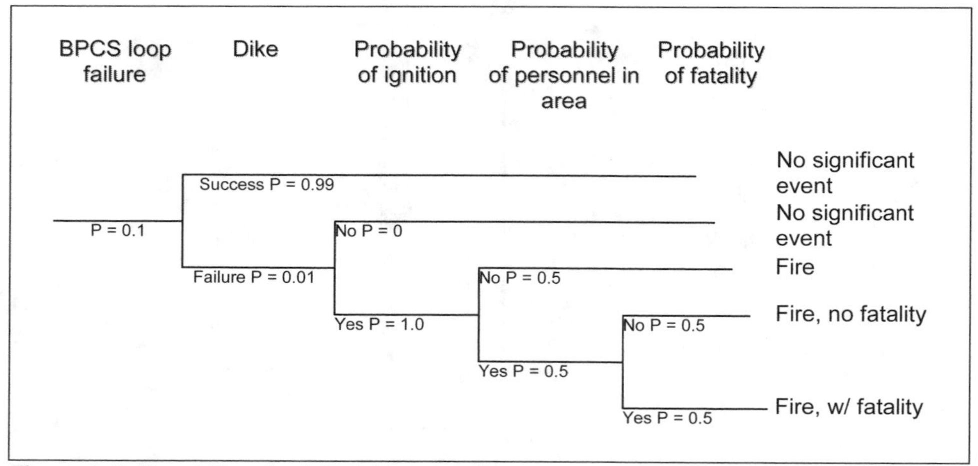

Figure 6-5: Event Tree for LOPA Example

Summary

Many industries have a need to evaluate and rank risk and match it to safety system performance. There are a variety of techniques that can be used ranging from qualitative to quantitative. All methods involve evaluating at least the two components of risk; probability and severity. No method is more accurate or better than another. All are valid techniques developed in different countries at different times.

The task of performing hazard and risk analyses and matching risk to safety system performance does not fall solely on the instrument or control system engineer. It's a task that must be carried out by a multi-disciplined team. Also, one determines the SIL (safety integrity level) for each safety instrumented function (SIF). It's incorrect to refer to the SIL of an entire process unit or piece of equipment.

LOPA Worksheet

Scenario Number:	**Equipment Number:**	**Scenario Title:** Hexane Surge Tank Overflow. Spill not contained by the dike.	
Date:	**Description**	**Probability**	**Frequency (per year)**
Consequence Description / Category	Release of hexane outside the dike due to tank overflow and failure of dike with potential for ignition and fatality		
Risk Tolerance Criteria (Category or Frequency)	Maximum tolerable risk of serious fire Maximum tolerable risk of fatal injury		$< 1 \times 10^{-4}$ $< 1 \times 10^{-5}$
Initiating Event (typically a frequency)	BPCS loop failure		1×10^{-1}
Enabling Event or Condition		N/A	
Conditional Modifiers (if applicable)			
	Probability of ignition	1	
	Probability of personnel in area	0.5	
	Probability of fatal injury	0.5	
	Other	N/A	
Frequency of Unmitigated Consequence			2.5×10^{-2}
Independent Protection Layers			
	Dike (existing)	1×10^{-2}	
Safeguards (non-IPLs)			
	Human action not an IPL as it depends upon BPCS generated alarm. (BPCS failure considered as initiating event.)		
Total PFD for all IPLs		1×10^{-2}	
Frequency of Mitigated Consequence			2.5×10^{-4}
Risk Tolerance Criteria Met? (Yes/No): No. SIF required			
Actions Required:	Add SIF with PFD of at least 4×10^{-2} (Risk Reduction Factor > 25) Responsible Group / Person: Engineering / J.Q. Public, by July 2005 Maintain dike as an IPL (inspection, maintenance, etc.)		
Notes:	Add action items to action tracking database		

Figure 6-6: Sample LOPA Example

References

1. US MIL-STD-882D-2000. *Standard Practice for System Safety.*
2. ANSI/ISA-84.00.01-2004, Parts 1-3 (IEC 61511-1 to 3 Mod). Application of Safety Instrumented Systems for the Process Industries.
3. *Guidelines for Safe Automation of Chemical Processes.* American Institute of Chemical Engineers - Center for Chemical Process Safety, 1993.
4. IEC 61508-1998. *Functional Safety of Electrical/Electronic/Programmable Electronic Safety-Related Systems.*
5. *Layer of Protection Analysis: Simplified Process Risk Assessment.* American Institute of Chemical Engineers - Center for Chemical Process Safety, 2001.
6. Smith, David J. *Reliability, Maintainability, and Risk: Practical Methods for Engineers.* 4th edition. Butterworth-Heinemann, 1993. (**Note:** 5th [1997] and 6th [2001] editions of this book are also available.)
7. Leveson, Nancy G. *Safeware - System Safety and Computers.* Addison-Wesley, 1995.
8. Marszal, Edward M. and Eric W. Scharpf. *Safety Integrity Level Selection: Systematic Methods Including Layer of Protection Analysis.* ISA, 2002.

Additional Background Material

1. Withers, John. *Major Industrial Hazards: Their Appraisal and Control.* Halsted Press, 1988.
2. Taylor, J. R. *Risk Analysis for Process Plant, Pipelines, and Transport.* E & FN Spon, 1994.
3. Cullen, Hon. Lord W. Douglas. *The Public Inquiry into the Piper Alpha Disaster*. Her Majesty's Stationery Office, 1990.

7

CHOOSING A TECHNOLOGY

Chapter Highlights

"If architects built buildings the way programmers wrote software, the first woodpecker that came along would destroy civilization."

Unknown

There are a number of technologies available for use in shutdown systems; pneumatic, electromechanical relays, solid state, and programmable logic controllers (PLCs). There is no one overall best system, just as there is no overall best car (vendor claims notwithstanding!). Each technology has advantages and disadvantages. It's not so much a matter of which is best, but rather which is most appropriate, based on factors such as budget, size, level of risk, complexity, flexibility, maintenance, interface and communication requirements, security, etc.

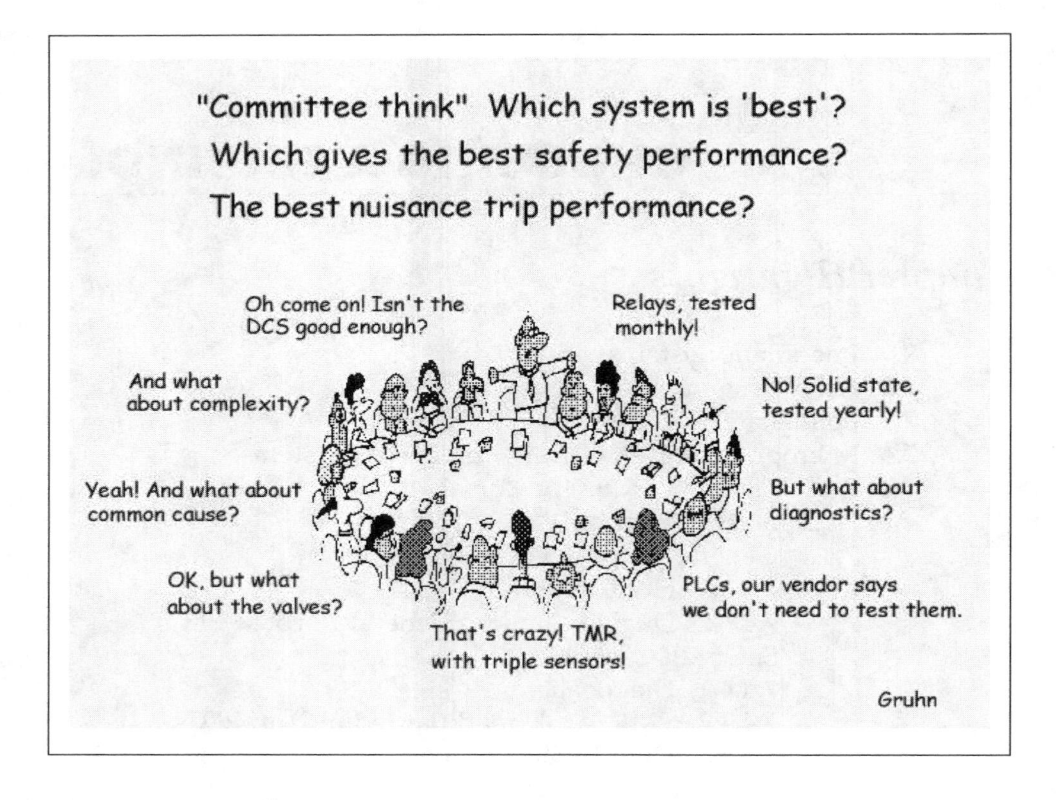

7.1 Pneumatic Systems

Pneumatic systems are still in use and are perfectly appropriate for certain applications. A very common application for pneumatic systems has been the offshore industry where systems must operate without electrical power. Pneumatic systems are relatively simple (assuming they're small) and relatively fail-safe. Fail-safe in this sense means that a failure or leak would usually result in the system depressurizing which would initiate a shutdown. Clean, dry gas is typically necessary. If desiccant dust from instrument air dryers or moisture from ineffective drying or filtering enters the system, small ports and vents utilized in the pneumatic circuits will be prone to plugging and sticking. This can render the circuits prone to more dangerous failures where the system may not function when required. Frequent operation and/or testing is usually necessary in order to prevent parts from sticking. Offshore operators in the U.S. Gulf of Mexico are required to test pneumatic safety systems on a monthly basis for this very reason. Pneumatic systems are typically used in small applications where there is a desire for simplicity, intrinsic safety and where electrical power is not available.

7.2 Relay Systems

Relay systems offer a number of advantages:

- *Simple* (at least when they're small)
- Low capital cost
- Immune to most forms of EMI/RFI interference
- Available in different voltage ranges
- *Fast response time* (they do not scan like a PLC)
- *No software* (which is still a concern for many)
- Fail-safe

Fail-safe means that the failure mode is known and predictable (usually with closed and energized circuits failing open and de-energized). However, nothing is 100% fail-safe. There are safety relays offering 99.9+% fail-safe operation.

Relay systems also have a number of disadvantages:

- *Nuisance trips:* Relay systems are typically non-redundant. This means that a failure of a single relay can result in a nuisance trip of the process. This can have a significant impact on overall operating costs.

- *Complexity of larger systems:* The larger a relay systems gets, the more unwieldy it gets. A 10 I/O (input and output) relay system is manageable. A 700 I/O relay system is not. An end user described to one of the authors certain problems they experienced with their 700 I/O relay system. One circuit consisted of a pneumatic sensor that went to an electric switch, that went to a PLC, that went to a satellite (honest!), that went to a DCS, that went to the relay panel, and worked its way out again. (This was obviously not your typical relay system!) The engineer said that between sensor and final element there were *57* signal handoffs! The author asked, "What's the likelihood that you think that's actually going to work when you need it?" The response was, "Zero. We know it won't work." Upon further questioning the engineer even admitted, "Everyone in this facility knows, 'Don't go near that panel!' The mere act of opening panel doors has caused nuisance shutdowns!"

- *Manual changes to wiring and drawings:* Any time logic changes are required with a hardwired system, wiring must be physically changed and drawings must be manually updated. Keeping drawings up-to-date requires strict discipline and enforcement of procedures. If you have a relay panel that is more than a decade old, try this simple test.

Obtain the engineering logic drawings and go out to the panel. Check to see if the two actually match. You may be in for a bit of a surprise! The problems associated with logic changes (i.e., changing wiring and keeping documentation up to date) were just some of the difficulties faced by the automotive industry. You can just imagine the thousands of relay panels they had for the production of all of the different automobiles, as well as the constant changes that were required. These were actually the primary issues that led to the development of the programmable logic controller (PLC)!

- *No serial communications:* Relay systems offer no form of communication to other systems (other than repeat contacts). Essentially, they're deaf, dumb, and blind.

- *No inherent means for testing or bypassing:* Relay systems do not incorporate any standard features for testing or performing bypasses. These features may be added, but the size, complexity, and cost of the panels increase significantly.

- *Discrete signals only:* Relay systems are based on discrete (on/off) logic signals. Traditionally, discrete input sensors (i.e., switches) were used. Analog (i.e., 4-20 mA) signals can be incorporated with the use of trip-amplifiers which provide a discrete output once an analog setpoint has been exceeded. Trip-amplifiers, however, do not offer the same inherent fail-safe characteristics as relays.

Relay systems are generally used for very small systems, typically less than 15 I/O. Relay systems are safe and can meet SIL 3 performance requirements, assuming the appropriate relays are used and the system is designed properly.

7.3 Solid-state Systems

Solid-state systems were designed to replace relays with smaller, lower power solid-state circuits (e.g., CMOS: complimentary metal oxide semiconductor), much as transistor radios replaced older, larger, vacuum tube designs. While still manufactured (only by two European vendors as of this writing), these systems are very specialized, are relatively expensive (compared to other choices), have limited applications, and as a result, are no longer very common.

Historically, there have been two different types of solid-state systems. One European design is referred to as "inherently fail-safe solid state." Similar to a relay, it has a known and predictable failure mode. Other systems could be categorized as "conventional" solid state, meaning they have more of a 50/50 failure mode split characteristic (50% safe, nuisance trip failures; 50% dangerous, fail-to-function failures).

Solid-state systems have a number of advantages:

- *Testing and bypass capabilities:* Solid-state systems built for safety generally include features for testing (e.g., pushbuttons and lights) and performing bypasses (e.g., keys).

- *Serial communication capabilities:* Some of the systems offer some form of serial communications to external computer based systems. This can be used to display system status, annunciate alarms, etc.

- *High speed:* These systems do not scan the I/O the way a PLC does. Therefore, these systems can respond faster than software-based systems (not considering things like input filter delays, etc.).

Solid-state systems have a number of disadvantages (many similar to those of relay systems):

- *Hardwired:* Solid-state systems are hardwired, similar to relays. Opening these panels and looking at all of the wiring and connections inside can be intimidating. Whenever logic changes are required, wiring and drawings must be manually updated.

- *Binary logic:* Solid-state systems perform the same sort of binary (on/off) logic as relays. They're not capable of analog or PID control, although most are now able to accommodate analog input signals. This naturally prohibits them from being used in applications that require math functions.

- *High cost:* These systems can be very expensive, sometimes even more expensive than triplicated PLCs. One might then ask, "Why use them at all?" The answer is simple. There are some who want more functionality than relays, but do *not* want a safety system that relies on software.

- *Non-redundant:* Like relays, solid-state systems are usually supplied in a non-redundant configuration. As with relays, single module failures can result in a nuisance trip to the process.

Whichever solid-state system one may consider, all of them offer similar construction (see Figure 7-1). There are individual modules for inputs, outputs, and logic functions. These systems do not use multiple channel input and output cards like PLCs. The modules must be wired into the logic configuration required. This may involve literally miles of wire in some systems. When you stop and consider all of the detail engineering and manufacturing required, it's no wonder these systems are so expensive.

The primary use for solid-state systems are small, high safety integrity level (SIL), applications. Two vendors have systems certified for use in SIL 4. One English document recommends the use of these systems (and not software-based systems) for small offshore applications with SIL 3 or greater requirements. [Ref. 1]

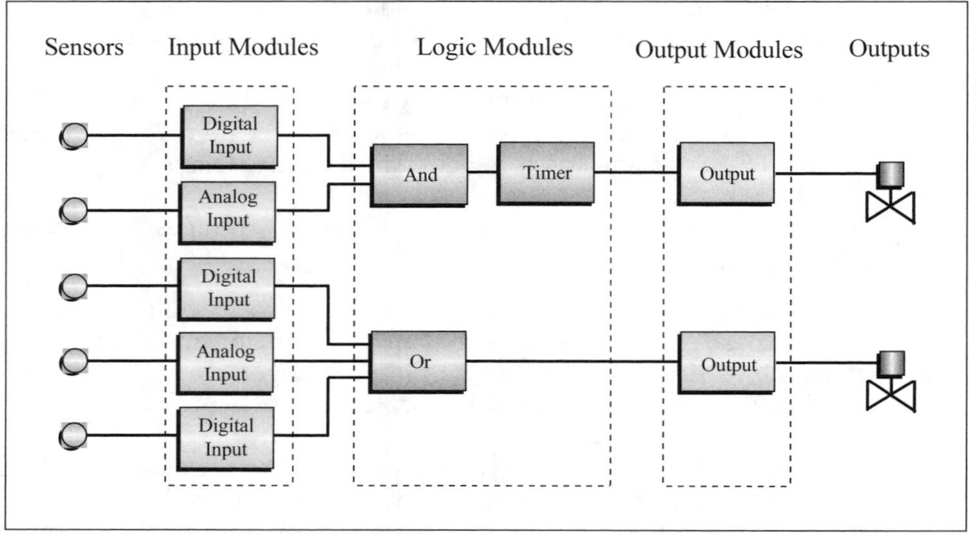

Figure 7-1: Typical Solid-state System

7.4 Microprocessor/PLC (Software-based) Systems

Software-based systems are utilized in the largest percentage of applications. There is no technological imperative, however, that says we *must* use microprocessor-based systems. Programmable logic controllers (PLCs) were originally designed to replace relays. Their use in safety systems would therefore seem inevitable, yet their application in safety requires close scrutiny.

Software-based systems offer a number of advantages:

- Reasonable cost
- Ease and flexibility in making changes
- Serial communications
- Graphic operator interfaces

- Self-documenting
- Small footprint

General-purpose PLCs, however, were *not* designed for use in critical safety applications. Most units do not have the diagnostic capabilities, fail-safe characteristics, or effective levels of redundancy for use beyond SIL 1 applications.

7.4.1 Flexibility: Cure or Curse?

Flexibility (i.e., the ease of making changes) offers benefits for some, but introduces potential problems. The ease of making changes will no doubt encourage actually *making* changes. Unless effective management of change (MOC) procedures are in place, this can easily lead to additional complexity and introduce new errors. Flexibility encourages redefinition of tasks late in the development phase, or even after installation, in order to overcome deficiencies found elsewhere in the system. John Shore phrased it well with, "Software is the place of afterthoughts." Flexibility also encourages premature construction. Few engineers would start to build a chemical plant before the designers had finished the detailed plans, yet this often happens with software-based systems. Flexibility also allows unique and sometimes unproven techniques to be used. Similarly, few engineers would design a complex system, such as a jet aircraft, after having only built a model.

Hardwired systems impose physical limitations on the design of a system. This helps control complexity. After all, you can only connect so many wires, timers, and relays in a given panel. In contrast, software has no corresponding physical limits. It's possible to build extremely complex software-based systems.

The level of automatic internal diagnostics with software-based systems varies considerably. Since most general-purpose PLCs were designed for active, dynamic control, they do not have a need for extensive diagnostics. Most failures are self-revealing in active control systems and diagnostics only increases their cost. This lack of diagnostics tends to be the primary weak link in these systems (as far as safety is concerned).

7.4.2 Software Issues

"And they looked upon the software and saw that it was good. But they just had to add this one other feature."

G. F. McCormick

Just as many would list software as the primary benefit of these systems, others view this dependence upon software as the primary *negative* consideration. This has to do with two areas: reliability and security. How many programmers, when they're through with a project, can walk into their boss's office, put their hands over their hearts and claim, "Boss, it's going to work every time, I guarantee it."? Also, most programmable systems offer very little in terms of access security.

According to one estimate, errors in the system's software (the software supplied with the system) can vary between one error in thirty lines at worst and one error in a thousand lines at best.

Michael Gondran [Ref. 2] quotes the following figures for the probability that there will be a significant error in the applications software of a typical microprocessor-based control system:

- Normal systems: 10^{-2} to 10^{-3}.
- To achieve 10^{-4} considerable effort is needed.
- To achieve 10^{-6} the additional effort has to be as great as the initial development effort.

Some vendors jokingly refer to software errors as "undocumented features." When software errors are found, letters are generally sent to all registered customers informing them of the issue. As you might imagine, it's generally *not* the vendors that discover most software problems. Most problems are discovered by users out in the field. They then inform the vendor who attempts to duplicate the problem. Some errors have caused nuisance trips. Some have prevented systems from working properly. A number of errors have been revealed when software was changed and downloaded to the system. Online changes (i.e., changes made to the controller while the process is still running) are generally not recommended.

Single failures in relay or solid-state systems only impact a single channel. Single failures in a PLC processor module, however, can impact an entire system. Therefore, many software-based systems utilize some form of redundancy, usually of the main processor. One issue associated with redundant systems is common cause. Common cause is a single stressor or fault that can make redundant items fail. Software is one such potential problem. If there's a bug in the software, the entire redundant system may not operate properly.

7.4.3 General-Purpose PLCs

The PLC was introduced in 1969 and was designed to replace relay control systems. It was inevitable that people would implement PLCs in safety systems. General-purpose PLCs offer a number of advantages, but they also have severe limitations when used in safety. Simply replacing relays with PLCs has actually caused some accidents. Just because it's a newer technology, doesn't necessarily mean its better in every regard.

7.4.3.1 Hot Backup PLCs

While many do use non-redundant PLCs for safety applications, many PLC systems are available with some form of redundancy. One of the most popular schemes is a hot backup system (see Figure 7-2). This system employs redundant CPUs—although only one is online at a time—and simplex I/O modules. Some systems have provisions for redundant I/O modules, others do not. The switching mechanism between the redundant CPUs can be accomplished a variety of different ways. There are a number of items worth pointing out about such systems.

In order for the system to switch to the back-up CPU, the system must first be able to detect a failure within the primary, or on-line, unit. Unfortunately, this does not happen 100% of the time because CPU diagnostics are unable to detect 100% of all possible failures. A general figure is about 80% diagnostic coverage, possibly getting as high as 95% if a number of additional features are implemented (e.g., watchdog timers). (Diagnostics are the topic of Section 7.4.5.1 in this text.) There have been reports of failures that were not detected, so the system did not switch. There have been reports of systems that were operating properly, yet the switch functioned when it was not supposed to, and the transfer to the backup unit was not a bumpless transfer and shut the process down. There have been reports where the system switched to the backup, only then to discover that the backup also had a failure, causing a system-wide shutdown.

Most PLC vendors assume, at least in their system modeling, that the switching is 100% effective. This is obviously an idealistic assumption as a variety of switch failures have been reported. One user reported pulling out the main processor in order to test the functionality of the switch. Nothing happened. The user then found that the switch needed to be programmed in order to work. Many PLC salespeople have privately reported that redundant (general-purpose) PLCs have more problems than non-redundant systems.

The real weak link, however, in most general purpose PLCs is the potential lack of diagnostics within the I/O modules. Some units literally have

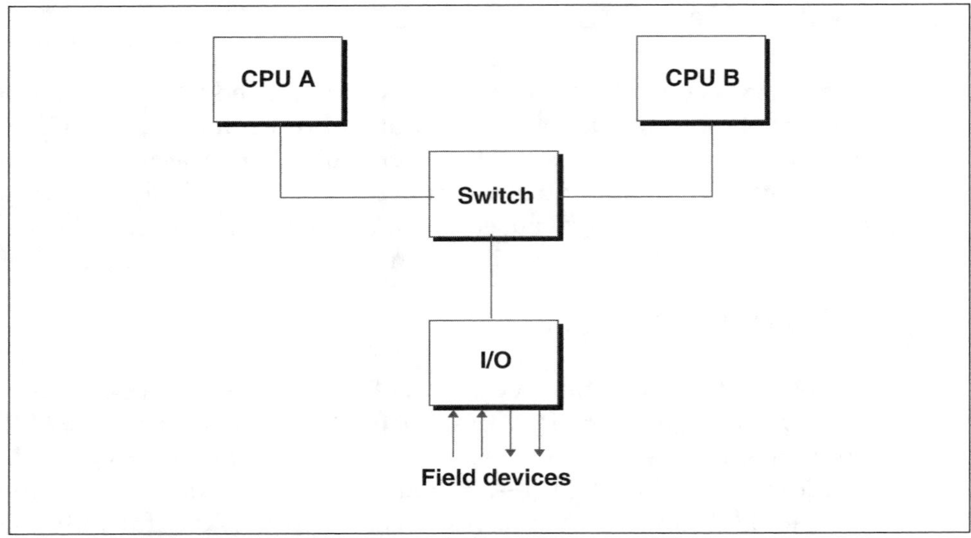

Figure 7-2: Hot Backup PLC System

no diagnostic capabilities at all. Users are *urged* to ask their vendors for this sort of information. Even ANSI/ISA-84.01 stated in 1996 that vendors were supposed to provide it. There are some units with relatively good diagnostics. When asked why some aren't using the systems with more thorough diagnostics, the typical response is, "They're too expensive." No kidding! You get what you pay for!

A number of rather frightening stories have been told of general purpose systems used in safety applications. Unfortunately, they're not widely publicized due to their sensitive nature.

For example, one of the authors knows an engineer who told how his company was considering the use of a PLC for safety. They put a demo system through a number of tests. One test was to take a running system, with all of the I/O energized, and remove the CPU chip *while the system was running*. They did just that and the system *did not respond* in any way! All of the I/O stayed energized and there were no alarms or indications of any kind of problem! He phrased it rather well, "We gave the system a lobotomy and it didn't even recognize anything happened!" His next sentence was, "We obviously didn't use that system."

A facility that had seven general-purpose systems for safety-related applications heard some troubling stories so the users went out and tested the units. (Many realize the need to test field devices, but don't believe that testing of the logic systems is necessary.) They essentially performed a site acceptance test and tickled each input, checking for the appropriate output response. Of the seven systems they tested, four did not respond properly! What's alarming is that no one knew anything was wrong until they actually went out and tested the systems. All the lights were green so they naturally assumed everything was OK.

The engineer of a corporate office reported that he sent out a memo to all of his companies' plant sites requesting that their PLCs used for safety be tested. The reports he got back indicated that between 30 and 60% of the systems did not respond properly! Again, this does not make headline news and is a very sensitive topic. No one presents a paper at an ISA conference saying "50% of our safety systems don't work. Boy, did *we* screw up!"

Another engineer reported that his E&C company designed a hot backup PLC system per the customer's specifications. They installed it, tested it, and said everything worked fine. They tested it a year later and said everything still worked fine. Then one day the PLC shut the plant down. When they checked the system, they found that the online unit failed, the system switched to the backup, but the backup CPU cable was never installed! The system ran for 1½ years and couldn't even detect or alarm that the backup wasn't even connected! The system was obviously not tested thoroughly at the integrator's facility, or during plant startup, or during maintenance. This would be an example of the 15% of errors found by the U.K. Health & Safety Executive classified as "design and implementation errors."

7.4.3.2 Dual Redundant General-Purpose PLCs

Some have taken off-the-shelf general-purpose PLC hardware and modified the system (i.e., customized it) in order to try and provide a higher level of diagnostic coverage. One such scheme involves the addition of I/O modules, loop back testing, and additional application programming (see Figure 7-3).

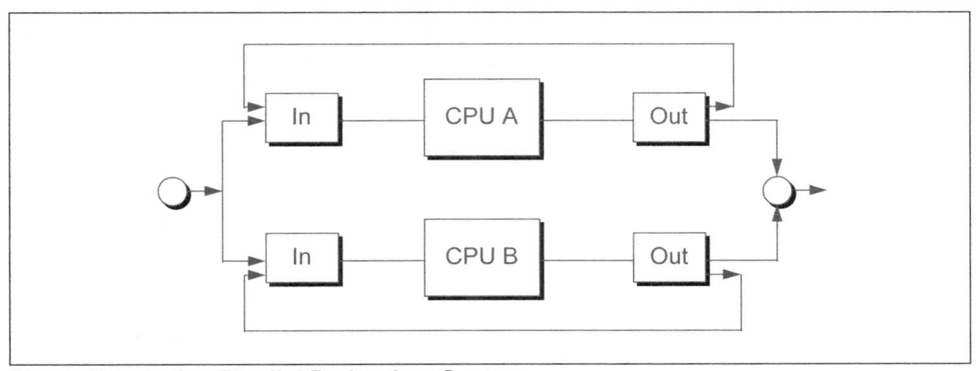

Figure 7-3: Active Parallel Redundant System

Unfortunately, systems of this type tend to be large and customized, and utilize extra programming which some consider confusing and difficult to maintain. When one considers the extra hardware, programming and engineering effort, this can become rather expensive. A word used by some (which others might be offended by) to describe such a system would be "kludge". One of the authors has known some who tried this approach and then said, "Never again!" This special dual approach might have made sense in the 1980s, but it's not very economical when compared against standard systems that are now available.

There's an inherent problem with any dual system of this type. When the two channels don't agree, how can you tell which one's right? Is one channel experiencing a safe failure? In this case you would *not* want to trip, so you might choose 2oo2 (two-out-of-two) voting. Or might one channel be non-responsive due to a dangerous failure? In this case, the alarm might be genuine and you *would* want to trip, so you might choose 1oo2 (one-out-of-two) voting. Again, which one's right? With most standard off-the-shelf systems, this is very difficult, if not impossible, to determine.

7.4.4 Safety PLCs

Certain users and vendors realized the limitations of general-purpose PLC systems in critical applications and went on to develop more appropriate systems. NASA led much of this research into fault tolerant computing back in the 1970s. The primary distinguishing features of safety PLCs are their level of diagnostics, implementation of redundancy, and independent certifications.

7.4.4.1 Diagnostics

Diagnostics are an important aspect of any safety system. Diagnostics are needed in order to detect dangerous failures that could prevent the system from responding to a demand. Many general-purpose PLCs do not incorporate thorough diagnostics. In fact, diagnostics are generally not required for active control systems where most failures are self-revealing.

The level of diagnostic coverage (i.e., the percentage of failures that can automatically be detected by the system) has a significant impact on safety performance. This will be covered further in Chapter 8 on evaluating system performance. The 1996 version of ANSI/ISA-84.01 stated that the logic solver supplier shall provide failure rate, failure mode, and diagnostic coverage information. How is this done?

The U.S. military was faced with similar problems and developed techniques to estimate failure rates and failure modes of electronic equipment.

One technique, referred to as failure modes and effects analysis (FMEA), has been modified slightly and used by control system vendors. The modified technique is referred to as failure modes, effects, and diagnostic analysis (FMEDA). It involves documenting all components in a system, how they may fail, what their failure rate is, and whether failures can automatically be detected by the system.

Figure 7-4 is a diagram of a PLC input circuit. Figure 7-5 represents the FMEDA of the circuit. The spreadsheet identifies each component in the circuit, all known failure modes of each device, the effect each failure would have on system operation (e.g., a nuisance trip, or a fail-to-function failure), the failure rate of each device in each failure mode, and whether each device and each failure mode can automatically be detected by the system. At the end of the study, the vendor can tell its customer the safe failure rate, dangerous failure rate, and level of diagnostics of the system. General-purpose systems typically have very low levels of diagnostics. What categorizes a safety PLC as such is its much higher level of automatic diagnostics.

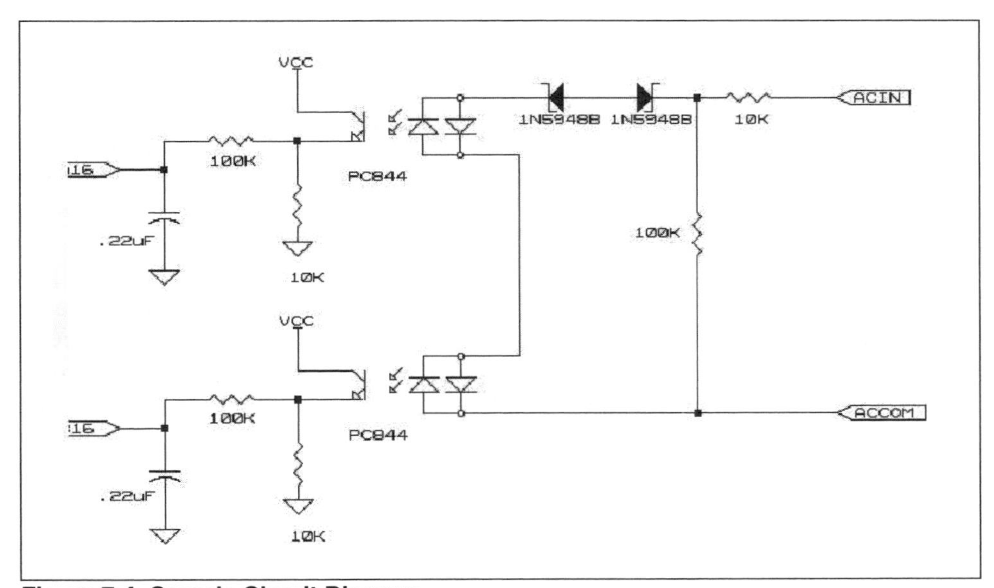

Figure 7-4: Sample Circuit Diagram

7.4.4.2 TMR Systems

Triple modular redundant (TMR) systems are essentially specialized triplicated PLCs (see Figure 7-6). The original research for these systems was funded by NASA in the 1970s. The systems that were developed commercially offer the same benefits listed above for PLCs, yet were specifically

Failure Modes and Effects Analysis					Failures/billion hours					Safe	Dangerous
Component	Mode	Effect		Criticality	FIT	Safe	Dang.	Det.	Diagnostic	Covered	Covered
R1 - 10K	short	Threshold shift	1	Safe	0.13	0.125	0	0		0	0
	open	open circuit	1	Safe	0.5	0.5	0	1	loose input pulse	0.5	0
R2 - 100K	short	short input	1	Safe	0.13	0.125	0	1	loose input pulse	0.125	0
	open	Threshold shift	1	Safe	0.5	0.5	0	0		0	0
D1	short	overvoltage	1	Safe	2	2	0	1	loose input pulse	2	0
	open	open circuit	1	Safe	5	5	0	1	loose input pulse	5	0
D2	short	overvoltage	1	Safe	2	2	0	1	loose input pulse	2	0
	open	open circuit	1	Safe	5	5	0	1	loose input pulse	5	0
OC1	led dim	no light	1	Safe	28	28	0	1	Comp. mismatch	28	0
	tran. short	read logic 1	0	Dang.	10	0	10	1	Comp. mismatch	0	10
	tran. open	read logic 0	1	Safe	6	6	0	1	Comp. mismatch	6	0
OC2	led dim	no light	1	Safe	28	28	0	1	Comp. mismatch	28	0
	tran. short	read logic 1	0	Dang.	10	0	10	1	Comp. mismatch	0	10
	tran. open	read logic 0	1	Safe	6	6	0	1	Comp. mismatch	6	0
R3 - 100K	short	loose filter	1	Safe	0.13	0.125	0	0		0	0
	open	input float high	0	Dang.	0.5	0	0.5	1	Comp. mismatch	0	0.5
R4 - 10K	short	read logic 0	1	Safe	0.13	0.125	0	1	Comp. mismatch	0.125	0
	open	read logic 1	0	Dang.	0.5	0	0.5	1	Comp. mismatch	0	0.5
R5 - 100K	short	loose filter	1	Safe	0.13	0.125	0	0		0	0
	open	input float high	0	Dang.	0.5	0	0.5	1	Comp. mismatch	0	0.5
R6 - 10K	short	read logic 0	1	Safe	0.13	0.125	0	1	Comp. mismatch	0.125	0
	open	read logic 1	0	Dang.	0.5	0	0.5	1	Comp. mismatch	0	0.5
C1	short	read logic 0	1	Safe	2	2	0	1	Comp. mismatch	2	0
	open	loose filter	1	Safe	0.5	0.5	0	0		0	0
C2	short	read logic 0	1	Safe	2	2	0	1	Comp. mismatch	2	0
	open	loose filter	1	Safe	0.5	0.5	0	0		0	0
					111	88.75	22			86.875	22
					Total	Safe	Dang.		Safe Coverage	0.9789	
					Failure Rates						
									Dangerous Coverage	1	

Figure 7-5: FMEDA of Sample Circuit Diagram

designed for safety applications and incorporate extensive redundancy and diagnostics. With triplicated circuits, such a system can survive single (and sometimes multiple) safe or dangerous component failures (hence the term "fault-tolerant"). These systems are suitable for use up to, and including, SIL 3 applications. This does *not* mean, however, that merely using such a logic box will provide a SIL 3 *system*. System performance also hinges on field devices, as we shall see.

The above diagram is generic. Some of the systems utilize quad redundant outputs. Some employ redundant circuits on one board (module), other utilize multiple modules.

In general, these systems do not require any additional overhead programming in order to accomplish diagnostics (although there are exceptions). The user only writes and loads one program, not three separate ones (although there are exceptions). Essentially, the triplication is designed to be transparent to the user.

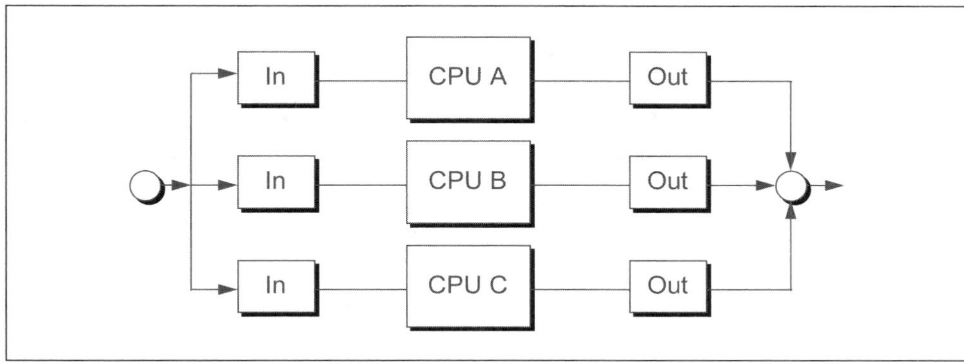

Figure 7-6: TMR System

7.4.4.3 Dual 1oo2D Systems

A number of dual systems have been developed since the late 1980s specifically for the process industry. The term 1oo2D (one-out-of-two with diagnostics) was coined in 1992 by William Goble in his ISA textbook [Ref. 3]. A number of the dual vendors quickly adopted the term. These systems are designed to be fault-tolerant to the same degree as triplicated systems. It takes two simultaneous safe failures to cause a nuisance trip and two simultaneous dangerous failures to fail to function. (Failure modes are covered in more detail in Chapter 8.) These systems are based on more recent concepts and technology and are certified by independent agencies such as TÜV and FM to the same safety levels as the triplicated systems. In addition to dual configurations, these vendors can also supply simplex (non-redundant) systems that are still certified for use in SIL 2, and in some cases even SIL 3, applications. An example of such a system is shown in Figure 7-7.

7.5 Issues Related to System Size

Most relay systems were small and distributed around the plant. Each process unit or piece of equipment typically had its own dedicated shutdown system. Just as computerized control systems in most plants became centralized, the shutdown systems also became more centralized. Rather than keeping multiple, small, hardwired systems, many users migrated to single, centralized, larger units. This was often the only way to afford the newer technology. However, the larger a system gets, the more complicated it may get. A small relay panel is relatively easy to manage and maintain. The same can't be said of a 500 I/O relay system. Hardwired systems were frequently replaced with physically smaller, easier to manage, software-based systems. This centralization, however, introduces a new set of problems. Single failures can now have a much wider impact.

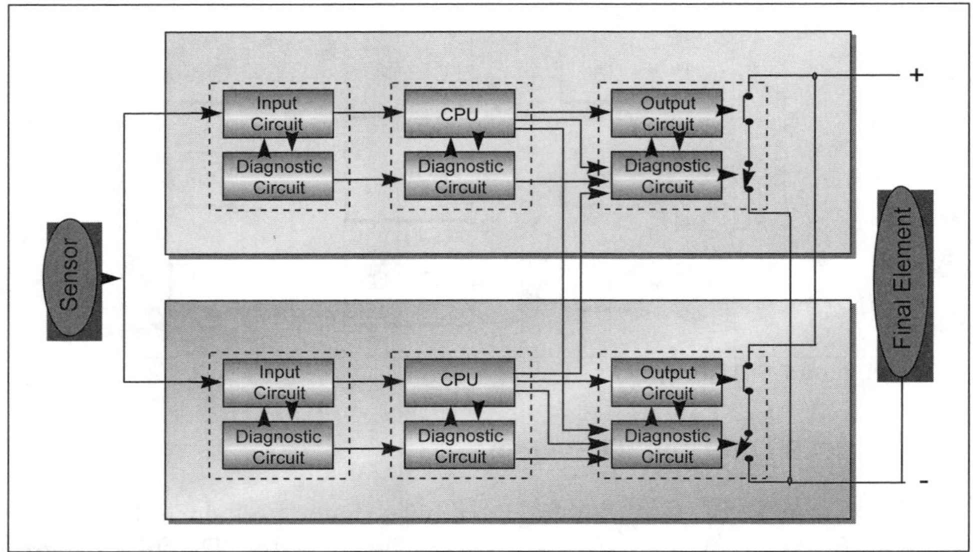

Figure 7-7: 1oo2D System

Single failures in small segregated systems will have a limited, localized impact only. Single failures in large centralized systems, however, may have a major impact. For example, if a non-redundant PLC is controlling 300 I/O and the CPU fails, control of all 300 I/O is lost. For this reason, many software-based systems are designed with some form of redundancy. The drawback of redundancy is that it usually results in additional system complexity. Some forms of redundancy (e.g., systems that switch to backup units) frequently introduce single points of failure and other common cause problems.

Size also has an impact in modeling of system performance. This topic will be covered in Chapter 8.

7.6 Issues Related to System Complexity

One way to deal with complexity is to break the design (e.g., software) into smaller pieces or modules. While this can reduce the complexity of the individual components, it increases the number of interfaces between all of the components. This can have a detrimental effect of actually increasing the overall complexity. It becomes difficult to analyze or even comprehend the many conditions that can arise through the interactions of all of the components.

In terms of safety instrumented systems, simpler is better. Instead of spending money for extra complexity, it may be better in the long run to spend money for simplicity. Simpler systems are easier to understand, analyze, design, build, test and maintain.

Complexity can be included when modeling the performance of a system. A functional or systematic failure rate can be incorporated in the calculations. This may be a design, programming, fabrication, installation, or maintenance error. While the exact number chosen may be rather subjective and difficult to substantiate, it's relatively easy to include in a model. Such an exercise is useful just to see what sort of an impact it has on overall system performance. Please refer to Chapter 8 for an example.

7.7 Communications with Other Systems

Most plants now have some form of centralized control system with graphical operator displays. Rather than having a separate set of displays for the safety instrumented system, there are benefits in having the SIS information available for display at the main control system consoles (e.g., lower cost, less space, learning only one system, etc.). This usually requires some form of serial communication link between systems.

Most computerized systems offer some form of serial communications. It's relatively easy to have the process control system read or poll the SIS and display information, such as the state of all the I/O, bypasses, alarms, etc. It's just as easy to have the process control system write to the SIS, and herein lies the potential danger.

Many control and safety systems are supplied from different manufacturers. In order for them to be able to communicate with each other, they must utilize a common communications protocol. One common protocol is Modbus. A limitation of Modbus is that it does not utilize or recognize alpha-numeric tag names. All variables must instead be referenced by numbers that are associated with memory address locations. Imagine a control and safety system communicating via Modbus. What might happen if changes are made to the database in the safety system? If variables are inserted in the middle of a list, rather than added at the end of a list, all of the remaining addresses will change. The control system will have no indication of the change and may end up communicating with all of the wrong variables as a result. This could naturally lead to serious problems. More recent and advanced communication methods, such as object linking and embedding for process control (OPC) should alleviate some of these problems.

Some companies allow writes to safety systems, some do not. There will no doubt be certain instances where the control system needs to write to the SIS. The standards do not prohibit this functionality. The primary consideration is that the control system must not be allowed to corrupt the memory of the SIS. This requires some form of control. Consider read-back testing to make sure only the intended variable has been altered. All writes to safety systems should be carefully considered and carefully controlled.

Safety fieldbuses, which are used to communicate between or within safety systems, are discussed in Section 9.8.

7.8 Certified vs. Prior Use

There are many vendors competing for business in every market, including safety. Products are often very diverse in design. Considering the complexity associated with many of the software-based logic systems, along with the conflicting stories told by many of the vendors, users faced a dilemma. How does one effectively evaluate so many diverse products? End users in the process industry specialize in processing oil, gas, and various chemicals. They're not in business to be specialists in evaluating safety systems. It's difficult for users to justify purchasing several different systems and evaluating them for several months just to see which one they like the best. This is especially true for brand new systems where there is no previous user experience. Enter the need for independent third-party assessments.

Competition between vendors is fierce in any industry, especially in safety. Dual vendors say their products are just as good as triplicated systems. Some vendors now offer quad redundant systems and naturally claim their systems are better than triplicated ones. Some vendors offer (seemingly) non-redundant systems that are suitable for use in SIL 3. Who is one to believe, and more importantly, why? Enter the need for independent third-party assessments.

One way for users to evaluate systems and for manufacturers to differentiate themselves in the safety market is with third-party certification. Organizations, such as Germany's TÜV, or FM in the U.S., are available to provide independent assessment against various standards. Prior to the release of the IEC 61508 standard in the late 1990s, TÜV used other German national standards.

The certification process is both time-consuming and expensive. Vendors must pass the cost along to their users. Certification was considered justifiable (at least by some) for complex and expensive logic systems. However, certification for simpler and less expensive field devices seemed questionable.

Users do not want to be *forced* to use certified equipment. The standards do *not* mandate the use of certified equipment. However, users still need some means of assuring themselves that equipment is suitable for use. The 1996 version of ANSI/ISA-84.01 referred to this concept as "user approved." The term used in the more recent IEC standards is "proven in use," or "prior use." In essence, the user must evaluate the vendor's quality program and have a large enough sample size in a similar environment to have meaningful failure rate data. This is generally considered an acceptable means of justifying the use of relatively simple field devices, but is considered extremely difficult to do for complex logic solvers.

Summary

There are a number of technologies available for use in safety instrumented systems: pneumatic, electromechanical relays, solid state, and PLCs (programmable logic controllers). There is no one overall best system. Each has advantages and disadvantages. The decision over which system may be best suited for an application depends upon many factors such as budget, size, level of risk, flexibility, complexity, maintenance, interface and communication requirements, security, etc.

Pneumatic systems are most suitable for small applications where there is a need for simplicity, intrinsic safety, and a lack of available electrical power.

Relay systems are fairly simple, relatively inexpensive to purchase, immune to most forms of EMI/RFI interference, and can be built for many different voltage ranges. They generally do not incorporate any form of interface or communications. Changes to logic require manually changing documentation. In general, relay systems are used for relatively small and simple applications.

Solid-state systems (i.e., hardwired systems that do not incorporate software) are still available. Several of these systems were built specifically for safety applications and include features for testing, bypasses, and communications. Logic changes still require manually updating documentation. These systems have fallen out of favor with many users due to their limited flexibility and high cost, along with the acceptance of software-based systems. They are still used in small, simple, high safety integrity level applications.

Software-based systems, ranging from general-purpose PLCs to specialized safety PLCs, offer software flexibility, self-documentation, communications, and higher level interfaces. Unfortunately, many general purpose systems were not designed specifically for safety and do not offer features (e.g., security, high levels of diagnostics, effective redundancy) that may be required. However, more recently designed safety PLCs were developed specifically for the more critical applications and have become firmly established in the process industries.

References

1. *Guidelines for Instrument-Based Protective Systems*. U.K. Offshore Operators Association, 1999.
2. Taylor, J. R. "Safety assessment of control systems - the impact of computer control." Israel Institute for Petroleum and Energy Conference on Process Safety Management held in Tel Aviv, Israel, October 1994.
3. Gondran, M. Launch meeting of the European Safety and Reliability Association held in Brussels, Belgium, October 1986.
4. Goble, W. M. *Evaluating Control Systems Reliability*. ISA, 1992.

8
INITIAL SYSTEM EVALUATION

Chapter Highlights

8.1 Things Are Not as Obvious as They May Seem
8.2 Why Systems Should be Analyzed Before They're Built
  8.2.1 Caveats
8.3 Where to Get Failure Rate Information
  8.3.1 Maintenance Records
  8.3.2 Vendor Records
  8.3.3 Third-party Databases
  8.3.4 Military Style Calculations
8.4 Failure Modes
  8.4.1 Safe/Dangerous Failures
  8.4.2 Detected/Undetected Failures
8.5 Metrics
  8.5.1 Failure Rate, MTBF, and Life
8.6 Degree of Modeling Accuracy
8.7 Modeling Methods
  8.7.1 Reliability Block Diagrams
  8.7.2 Fault Trees
  8.7.3 Markov Models
8.8 The Real Impact of Redundancy
8.9 Basic Formulas
  8.9.1 Impact Due to Manual Test Duration
8.10 Analysis of a Relay System
8.11 Analysis of a Non-redundant PLC System
8.12 Analysis of a TMR System
  8.12.1 Common Cause
8.13 Field Devices
  8.13.1 Partial Stroking of Valves
8.14 Fault Tolerance Requirements
8.15 Sample SIS Design Cookbook
8.16 Engineering Tools Available for Analyzing System Performance
Summary
References

"There's always an easy solution to every human problem, neat, plausible... and wrong."

— *H. L. Menken*

8.1 Things Are Not as Obvious as They May Seem

If it were intuitively obvious which system was most appropriate for a particular application, then there would be no need for this book or any of the design standards. The problem is that things are *not* as intuitively obvious as they may seem. Dual is *not* always better than simplex, and triple is *not* always better than dual. Consider the nine choices presented in Table 8-1.

Let's assume at this point that all nine different cases shown in the table have software-based logic systems (programmable electronic system — PES). Let's not even concern ourselves with relay and solid state systems at this point.

Table 8-1: Which system is "best"? Which gives the fewest nuisance trips? Which gives the best safety performance?

Sensors	PES Logic	Diagnostic Coverage	Common Cause	Outputs	Test Interval
Single	Single	99.9%	N/A	Single	Monthly
Dual	Single	99%	N/A	Dual	Quarterly
Triple	Single	90%	N/A	Dual	Yearly
Single	Dual	99%	0.1%	Single	Monthly
Dual	Dual	90%	1%	Dual	Quarterly
Triple	Dual	80%	10%	Dual	Yearly
Single	Triple	99%	0.1%	Single	Monthly
Dual	Triple	90%	1%	Single	Quarterly
Triple	Triple	80%	10%	Single	Yearly

First, remember that safety instrumented systems can fail two ways. They may suffer nuisance trips and shut the plant down when nothing is actually wrong. They may also fail to function when actually required. Could one system be the best in both modes? If a system is good in one mode, would it necessarily be bad in the other?

A chain is only as strong as the weakest link. Should the sensors be redundant? Dual redundant devices can be configured at least two ways. One-out-of-two (1oo2) voting would cause a trip if one sensor were to go into alarm. Two-out-of-two (2oo2) voting would only trip if both sensors were to go into alarm.

Should the logic solver be redundant? There are at least four different ways of configuring dual logic solvers (1oo2, 2oo2, 1oo2D, or hot backup). What about triplicated?

Diagnostic coverage means what percentage of failures can be detected automatically by the system. No device has 100% diagnostic coverage. Would a non-redundant system with 99.9% diagnostic coverage be better than a triplicated system with 80% coverage?

Common cause means a single stressor or fault which impacts an entire redundant system. Typical examples are external environmental stresses to the system such as heat, vibration, over-voltage, etc. One method of quantifying common cause is referred to as the Beta factor. This represents the percentage of failures identified in one "leg" or "slice" of the system that might impact multiple channels at once. For example, if a redundant system has a 1% Beta factor, it means that of all the failures identified, 1%

of them might hit multiple channels at the same time and make the entire system fail. Is 1% enough to worry about? What about 10%?

Should the final elements, typically valves, be redundant? In parallel or in series? Redundant valves, especially large ones, will obviously be very expensive, but are they important? Triplicated outputs are not shown in the table. Triplication usually, although not always, refers to two-out-of-three voting. It's not possible to connect three valves and get two-out-of-three voting. However, three valves in series *have* been implemented.

How often should the entire system be manually tested? Testing is required because no system offers 100% diagnostics and can't detect all faults on its own. Should everything be tested at the same interval, or might different portions be tested at different intervals? Do redundant systems need to be tested more often, or less often, than non-redundant systems?

Go back to Table 8-1. Add a column on the left labeled "nuisance trips". Add a column on the right labeled "safety". Rank each row with the numbers 1 through 9, best through worst. Go ahead, do it!

How confident are you of your answers? Do you believe that your gut feel, intuition, or experience will give you the same answer as someone else?

Various standards groups met for *years* trying to resolve these issues. If all of this were easy, the groups would have been done a *long* time ago! Consider the ISA SP84 committee alone. There were over 300 members on the committee mailing list representing about every interest group imaginable (e.g., solid state, PLC, safety PLC, and DCS vendors; end users; contractors; integrators; consultants; certification agencies, etc.). How can such a diverse group reach agreement on a topic as complicated and controversial as this? How can one peer past all the hype?

Intuition may be fine for some things, but not others. Jet aircraft are not built by gut feel. Bridges are not built by trial and error, at least not any more. Nuclear power plants are not built by intuition. If you were to ask the chief engineer of the Boeing 777 why they used a particular size engine, how comfortable would you feel if their response was, "Well, we weren't sure... but that's what our vendor recommended."

8.2 Why Systems Should be Analyzed *Before* They're Built

Ideally, would you rather perform a HAZard and OPerability study (HAZOP) on a plant *before* you build it, or *afterwards*? The obvious answer is before, but not everyone who is asked this question realizes the real reason *why*. It's *cheaper* to redesign the plant on paper. The alternative would be to rebuild the plant after the fact. The same applies to safety systems.

As described above in Section 8.1, things are not as intuitively obvious as one may wish. Deciding which system is appropriate for a given application is not always a simple matter. It's therefore important to be able to analyze systems in a *quantitative* manner. While quantitative analyses may be imprecise (as will be stressed shortly), they are nevertheless a valuable exercise for the following reasons:

- They provide an early indication of a system's potential to meet the design requirements.
- They enable lifecycle cost comparisons.
- They enable one to determine the weak link in the system (and fix it, if necessary).
- They allow an "apples to apples" comparison between different offerings.

8.2.1 Caveats

"There are lies, there are damn lies, and then there's statistics."

— *M. Twain*

Simple models may be calculated and solved by hand. As more factors are accounted for, however, manual methods become rather unwieldy. It's possible to develop spreadsheets or other computer programs to automate the process. A major drawback of some models is often not what they include, but what they do *not* include. One can model a triplicated system according to one vendor's optimistic assumptions, and then model it with a more realistic set of assumptions, and change the answer by four orders of magnitude! It's not always the *accuracy* of the model that matters, it's more often the *assumptions* that go into it. Computers are known for their speed, not their intelligence.

"Computer models can predict performance with great speed and precision, yet they can also be completely wrong!"

— *Unknown*

When the Boeing 777 was being designed, there were two different factions regarding the modeling and testing of the engines. One group felt their computer models were so good that testing a real engine was unnecessary. Another group felt that actual testing was essential. The latter group eventually won the argument. Two engines on one side of a 747 were removed and a new single engine intended for the 777, which was twice as powerful as the earlier engines, was installed in their place. The first time the plane took off, the new engine flamed out. That particular flaw was not revealed in the computer models. Always keep in mind that models are just that—models—they are *not* reality!

"However neutral computers may be, they can never produce an objective answer from highly subjective data."

— *Imperato & Mitchell*

One needs to apply a bit of common sense in all modeling. For example, if two events each have a probability of 10^{-6}, a simplistic approach would be to say that the possibility of the two events happening *simultaneously* would be 10^{-12}. Low numbers such as this simply mean that the system will most likely fail in a way not even considered by the original designers. Absolutely absurd risk estimates based on failure rate data are not uncommon. [Ref. 2] The current record is an estimate of a nuclear weapon system where an event probability was estimated at 2.8×10^{-397}. Considering that the probability of being struck by a falling plane is roughly 10^{-8}, the absurdity of some computations becomes obvious. Kletz phrases it well and rather bluntly, "One wonder how scientifically trained people can accept such garbage." [Ref. 1]

"Reliability models are like captured foreign spies, if you torture them long enough, they'll tell you anything."

— *P. Gruhn*

Kletz also pointed out that "time is usually better spent looking for all the sources of hazard, than in quantifying with ever greater precision those we have already found." For example, in the space program, where quantitative fault tree analysis and failure modes and effects analysis were used extensively, almost 35% of actual in-flight malfunctions had *not* been identified. [Ref. 2]

8.3 Where to Get Failure Rate Information

In order to analyze and predict the performance of a system, one needs performance (i.e., failure rate) data of all the components. Where does one get this sort of information?

8.3.1 Maintenance Records

Hopefully, each facility has maintenance records indicating failure rates of safety related devices in the plant. The process safety management regulation (29 CFR 1910.119, Appendix C, Part 9) states that users *should* be keeping this sort of data. Many plants do in fact have the information, although it may not initially be collated in a usable fashion. Experienced technicians generally have a very good idea of how often they have to repair and/or replace different equipment.

Plant maintenance records are the *best* potential source of data. They're most representative of the application environment, maintenance practices, and vendor of concern, although obtaining statistically significant sample sizes are important.

But if one doesn't have such records, then what?

8.3.2 Vendor Records

One could ask vendors, but feedback from those who have tried is not very encouraging. Note that the 1996 version of ANSI/ISA-84.01 stated the logic solver supplier was supposed to provide such data. Even if the vendor *does* provide data, it's important to ask how *they* got it. Is it based on field returns or indoor laboratory experiments? How many users of a ten-year-old piece of equipment (e.g., a transmitter or solenoid valve) actually send failed devices back to the factory?

One PLC vendor stated that their failure rate data assumed that approximately 25% of units shipped were in storage (i.e., not in service) and that the remaining items were in use two shifts a day, five days a week. Not many safety systems in the process industry operate under those assumptions.

Field returns represent the *worst* potential source of data. Few users send in items past the warranty period; therefore, the data compiled is usually not statistically significant.

Vendor failure rate data, especially for field devices, generally does not take into consideration process or environmental factors. A device will most likely fail sooner in a dirty, corrosive service than in a clean service. Failures such as plugged sensing lines are generally not accounted for in vendor data (as they may consider it outside the scope of their device).

8.3.3 Third-party Databases

There are commercially available databases from the offshore, chemical, and nuclear industries, as well as general data. This information has been compiled and made available for everyone's use in the form of books and electronic databases. [Ref. 5 to 10] These databases are readily available, although not always cheap. They represent the *second best* potential source of data. They're based on actual field experience from different users, but are generic in nature. They do not differentiate between vendors or account for different application environments. Specialized safety system modeling programs also have failure databases built in. [Ref. 16 and 17]

But what about equipment that's brand new, or hasn't been built yet, and does not yet have a history of operating in the field?

8.3.4 Military Style Calculations

The military was faced with this problem decades ago. When a nuclear submarine receives the launch code, one wants to know the likelihood of the communications system working properly using some measure other than "high." The military developed a technique for predicting failure rates of electronic components (MIL-HDBK 217). This handbook has gone through numerous revisions and has long been the subject of much controversy. All who use it are aware that it tends to give pessimistic answers, sometimes by orders of magnitude. That does not mean, however, that it shouldn't be used, just that it should be used with caution. It still provides an excellent yardstick for comparing systems, even if the absolute answers are questionable. The publishers of the document are aware of this and state the following,

"... a reliability prediction should never be assumed to represent the expected field reliability as measured by the user..." (MIL HDBK 217F, Paragraph 3.3)

Many vendors now provide failure rate data using the military style calculation. Performing the calculations is even a requirement for some of the safety certifications. However, the calculations may not account for all factors or failures that users may experience out in the real world. For example, one sensor vendor published a mean time between failure (MTBF) of 450 years. Data compiled by users is typically an order of magnitude lower (e.g., 45 years). The difference may simply be due to the vendor not considering all possible application considerations and failure modes in their analysis (e.g., a plugged sensing line, installation issues, etc.), whereas the user database may have. Caution and common sense are warranted. Just because a vendor gives you a number doesn't mean the number is valid or realistic.

8.4 Failure Modes

Many people have replaced relay systems with programmable logic controllers (PLCs). Their typical response as to why they did so is usually, "That's what a PLC was designed for—to replace relays." How a system *operates* is certainly important. Obviously a PLC can do everything a relay system could, and a whole lot more (e.g., timers, math functions, etc.). However, the main concern for a safety system should not be so much how the system *operates*, but rather how the system *fails*. This concept is so simple that it's often overlooked. This is the underlying reason why dormant safety systems differ from active control systems and why safety instrumented systems have unique design considerations.

8.4.1 Safe/Dangerous Failures

It may be somewhat of a simplification, but safety systems are considered to fail in *two* ways. First, systems may initiate nuisance trips. They may shut the plant down when nothing is actually wrong. An example would be a closed and energized relay that just pops open. People have given these type of failures many different names; overt, revealed, initiating, fail-safe, etc. The term used in the standards for this type of failure is "safe failure". Granted, there's nothing "safe" about a nuisance trip, therefore many don't like this term. Safe failures result in plant shutdowns; therefore they tend to be costly in terms of lost production downtime. People want to avoid safe failure primarily for economic reasons. When systems suffer too many safe failures like this, people tend to lose confidence in them, and the systems may be bypassed as a result. Never forget that the availability of a system in bypass is zero. Accidents have happened because sensors or systems were placed in bypass while the process was allowed to run. One hotel operator went to jail for manslaughter for disabling a fire detection system which resulted in the death of a dozen guests. The implication of the 1992 OSHA process safety management regulation is that individual people may now be held criminally liable for such actions.

One must not forget, however, that safety systems may also suffer failures that will prevent them from responding to a true demand. You could think of these as "I'm stuck and I'm not going to work when you need me" type failures. Some have called these covert, hidden, inhibiting, or fail-danger faults. The standards now refer to this type of failure as a "dangerous failure". If a system fails in this manner (e.g., the relay contacts are welded shut) it would be *potentially* dangerous. A closed relay contact failing closed may not immediately result in a hazardous event. However, if there were a demand *at the same time* as such a failure, the system would not respond. The only way to find these failures—before it's too late—is to *test* for them. Many people simply don't understand the need for this. Remember, however, that safety instrumented systems are dormant, or passive. Not all failures are inherently revealed. Normally open valves that are stuck open do not indicate a problem. Normally energized circuits that fail short-circuited do not indicate a problem. Dormant PLCs that are stuck in endless loops do not indicate a problem. Sensors with plugged impulse lines do not always indicate a problem. Unfortunately, many general purpose systems do not have effective diagnostics in order to detect such failures. Vendors rarely discuss potential weaknesses such as this.

Examples of dangerous, hidden failures can be somewhat embarrassing and are therefore often "buried." One example was a satellite manufacturer that used a crane to lift a $200,000 satellite into its shipping container. When they raised the satellite, the crane was unable to stop the upward motion. When the satellite reached the top of the crane the cable snapped and the satellite crashed to the floor. The relay responsible for controlling the upward motion was replaced earlier with one rated for only 10% of the actual load, so the contacts welded. The crane was leased and was not tested before use.

8.4.2 Detected/Undetected Failures

Failures can be categorized not only as safe and dangerous, but as detected and undetected (see Figure 8.1). Safe failures are shown in the upper half of the diagram. Dangerous failures are shown in the lower half. Detected failures are shown in the left half. Undetected failures are shown in the right half. The portions of the pie chart are shown as equal sizes merely for simplicity. Detected failures have to do with automatic diagnostics. In other words, can the system automatically detect when a failure occurs. One measure for the level of diagnostics is referred to as the "coverage factor." Figure 8-1 shows a safe failure mode split of 50%, and a 50% diagnostic coverage factor. The Greek letter λ is used to express failure rate, the number of failures per unit of time. This term will be discussed in more detail shortly.

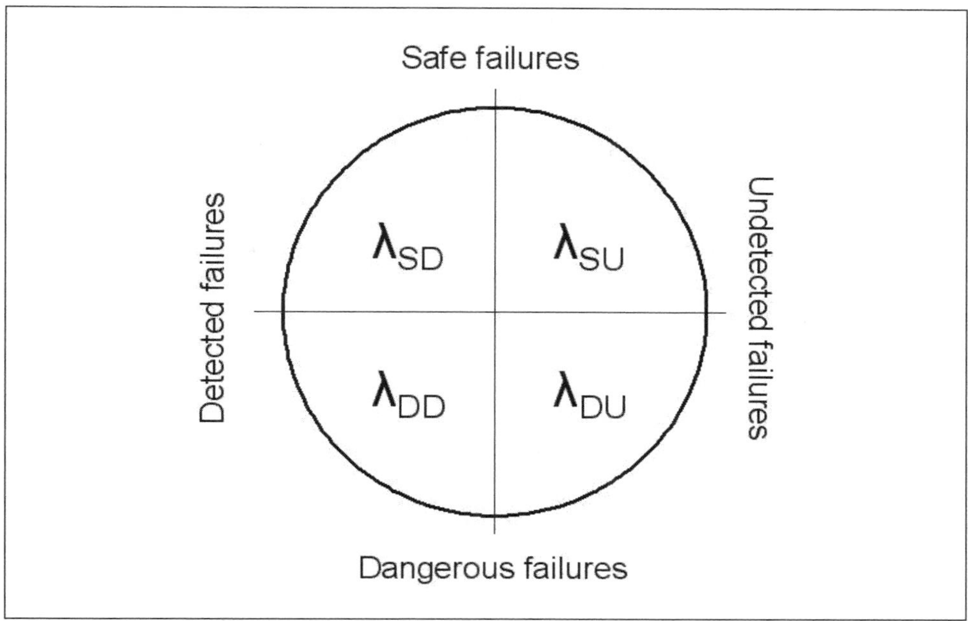

Figure 8-1: Failure Modes

8.5 Metrics

In order to measure and compare the performance of different systems, one needs a common frame of reference, a common *understandable* set of terms. A number of different performance terms have been used over the years, such as availability and reliability. Unfortunately, these seemingly trivial terms have caused problems. If you ask four people what they mean, you'll likely get four different answers. The main reason stems from the two different failure modes of safety systems discussed earlier. If there are two failure modes, there should be *two* different performance terms, one for *each* failure mode. How can one term, such as availability, be used to describe the performance of *two* different failure modes? If someone says a valve fails once every 10 years, what does that actually mean? What if 10% of the failures are "safe" (e.g., fail closed) and 90% are "dangerous" (e.g., fail stuck)? What if the numbers are switched? What if they're both the same? One overall number simply doesn't tell you enough.

Another reason the term availability causes confusion is the typical range of numbers encountered. Anything over 99% sounds impressive. Whenever PLC vendors give performance figures, it always ends up being a virtually endless string of nines. Stop and consider whether there's a significant difference between 99% and 99.99% availability. It's less than 1%, right? True, but the numbers also differ by two orders of magnitude! It can be confusing!

In terms of safety, some prefer to the compliment of availability—unavailability—also called the probability of failure on demand (PFD). Unfortunately, these numbers are usually so small that one needs to use scientific notation. This makes the numbers difficult to relate to for some. The purpose of this exercise is to give management meaningful information so they can make intelligent decisions. Will telling your supervisor, "Hey boss, our safety system has a PFD of 2.3 x 10^{-4}!" be meaningful or helpful? No doubt the manager's first response will be, "Uh, is that good?"

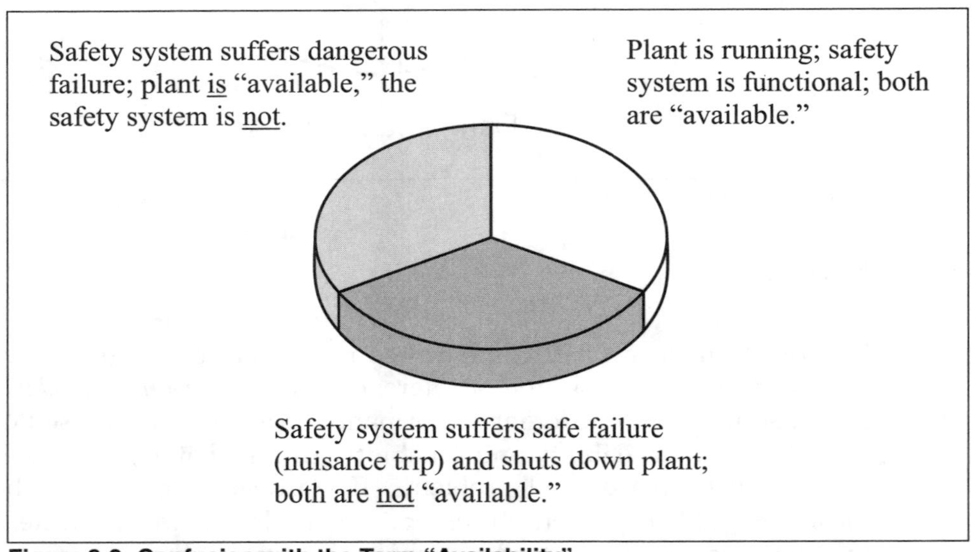

Figure 8-2: Confusion with the Term "Availability"

Consider the pie chart shown in Figure 8-2. The upper right third is where the safety system is performing properly and the plant is up and producing product. In this case, the SIS and the plant are both available. Using the term "availability" would seem to make sense.

In the lower section the SIS has suffered a nuisance trip and shut the plant down. In this case both are unavailable. (Granted, you do have to get the SIS up and running before you can get the plant started, so the two are not exactly the same, but that's getting picky.) Using the term "availability" (or "unavailability") would still seem to make sense.

However, the case most people are not even aware of is the upper left portion where the SIS is non-functional (e.g., a PLC stuck in an endless loop, outputs energized and stuck on, etc.), so the SIS is *not* available, but the plant *is* still up and running. This sort of system failure does *not* shut the plant down. The plant *is* available and still making product and oblivious to the fact that the SIS will not function when needed. In this case, the "availability" of the two systems (the SIS and the plant) is clearly *not* the same.

This also shows that you shouldn't combine the two different measures of performance with one term (such as an overall availability) because the impact of each failure is completely different. It hardly makes sense to take an average of 99.999% in the safe mode, and 94.6% in the dangerous mode. What would the final answer tell you? Absolutely nothing.

The generic term availability simply means uptime divided by total time. What might an availability of 99.99% actually mean? The system could have a nuisance trip once a month and be down for only 4.3 minutes. It could trip once a year and be down for 53 minutes. It could trip once every 10 years and be down for 8.8 hours. All work out to 99.99%. This is not the sort of information most users really want to know.

A better term for nuisance trips performance is mean time to failure, spurious ($MTTF^{spurious}$). This term was introduced in one of the ISA SP84 technical reports. [Ref. 11] In other words, will the system cause a nuisance trip, on average, once every 5 years, 50 years, or 500 years? Users know how long their particular process will be down when there is a nuisance trip. They just want to know how *often* it's going to happen.

A better term for dangerous system performance is risk reduction factor (RRF). This is the reciprocal of probability of failure on demand (PFD). The difference between 0.1 and 0.001 is difficult to visualize or appreciate. The difference between 10 and 1,000 however, is obvious.

8.5.1 Failure Rate, MTBF, and Life

Failure rate is defined as the number of failures per unit of time. The Greek letter λ is typically used to represent failure rate. The bath tub curve (Figure 8-3) illustrates that failure rates are not fixed over the entire life of a device. The curve shown is generally accepted for electronic devices. Software and mechanical devices tend to have slightly different curves. The left portion of the curve shows the impact of "infant mortality" failures. In other words, defective components die young. The right portion of the curve shows "wearout" failures. A constant failure rate is generally assumed for most devices. This is represented by the middle, flat portion of the curve. This simple assumption is still a topic of heated debate in the reliability community. The assumption at least tends to simplify the math involved. One may argue that devices such as valves do not obey the curve. The longer a valve is left in one position, the more prone it may be to sticking. This may be true, but until the industry comes up with more accurate models and data, the simplification will have to suffice. Since most assessments are generally only dealing in order of magnitude accuracy, this is usually not a problem.

Databases typically list failure rates expressed in units of failures per million hours. A reliability engineer may be comfortable relating to 1.5 failures per million hours, but not many others are. Most prefer the reciprocal of failure rate, mean time between failure (MTBF) with units of years. Knowing the two terms are merely reciprocals of each other, and that there are 8,760 hours in a year, a failure rate of 1.5 E-6 works out to a MTBF of 76 years.

Some argue over the difference between mean time to failure (MTTF) and mean time between failure (MTBF). Traditionally, MTTF is used for replaceable components (e.g., resistors), and MTBF is used for repairable systems (e.g., aircraft). MTBF can also be considered the combination of MTTF and mean time to repair (MTTR). Considering that MTBF values are typically measured in years, whereas MTTR values are typically measured in hours, one can see that the difference between MTTF and MTBF is usually not worth arguing about. For the sake of consistency MTBF is used in this chapter.

Many consider MTBF and "life" to be the same. For example, if a device has an MTBF of 30 years, many assume that the device lasts an average of 30 years and then fails. This is not true. An MTBF of 3,000 years may be a perfectly valid number for a device, even though no one expects the device to last for 3,000 years. The number merely means that out of a total of 3,000 devices, one device fails within a one year period. It's merely a statistical average. One can't predict *which* device or *when* within the time period that it will fail. A classic example to illustrate how MTBF and life are not the same is a match. When using dry matches and the proper technique, there should be few failures. Therefore, the failure rate (failures per unit time) will be low. If the failure rate is low, the reciprocal, MTBF, will be large, perhaps several minutes in the case of a match. But a match only burns, or has a life, of a few seconds.

8.6 Degree of Modeling Accuracy

All reliability analyses are based on failure rate data. It must be recognized that such data is highly variable. Allegedly identical components operating under supposedly identical environmental and operating conditions

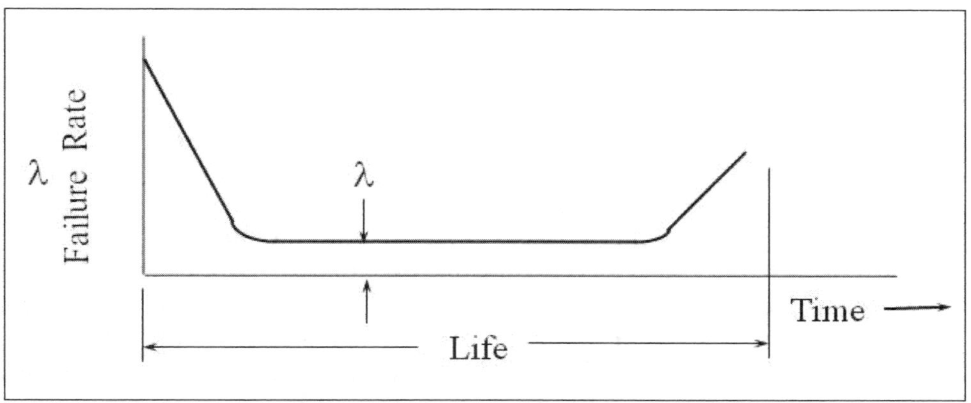

Figure 8-3: Bathtub Curve Illustrating Life, Failure Rate and MTBF (1/λ)

are not realistic assumptions. For a given level of detail, the apparent precision offered by certain modeling methods is not compatible with the accuracy of the failure rate data. As a result, it may be concluded that simplified assessments and the use of relatively simple models will suffice. More accurate predictions can be both misleading, as well as a waste of time, money, and effort. In any engineering discipline, the ability to recognize the degree of accuracy required is of the essence. Since reliability parameters are of wide tolerance, judgments must be made on one-, or at best two-, figure accuracy. Benefits are obtained from the judgment and subsequent follow up action, not from refining the calculations. [Ref. 4] Simplifications and approximations are useful when they reduce complexity and allow a model to become understandable. [Ref. 14] For example, if a simple model indicates the risk reduction factor of a system is 55, there's little point in spending an order of magnitude more time and effort, or hiring an outside consultant, to develop a model that indicates the risk reduction factor is 60. Both answers indicate the middle of the SIL 1 range (i.e., a risk reduction factor between 10 and 100).

8.7 Modeling Methods

"Managing lots of data begins with some form of simplification."

— *Megill*

"The less we understand a phenomenon, the more variables we require to explain it."

— *L. Branscomb*

There are a number of methods available for estimating the performance of systems. Some of the more common are reliability block diagrams, fault trees, and Markov models.

8.7.1 Reliability Block Diagrams

Reliability block diagrams (RBDs) are just that, *diagrams*. Their usefulness is in helping to clarify system configuration and operation. Figure 8-4 is an example. In this case the system would fail if A, B, or G failed. Items C and D, as well as E and F, are redundant. The system would only fail if both items C and D, or both E and F, failed at the same time. The formulas generally associated with reliability block diagrams simply involve adding or multiplying probabilities of the blocks. With the assumption that probabilities are small, one would add the probabilities of items in series (A, B and G). One would multiply the probabilities of items in parallel (C and D and E and F). In general, block diagrams and their associated math do not handle time dependent variables such as repair times, test intervals, diagnostics, and the more complex redundant systems. However, more complex formulas can be used to calculate failure probabilities of each block.

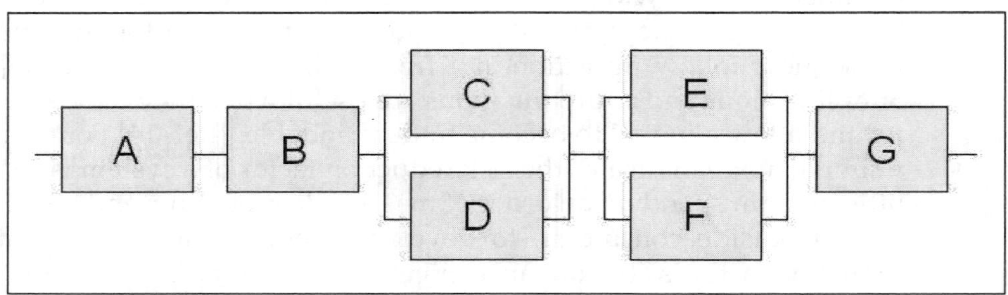

Figure 8-4: Reliability Block Diagram

8.7.2 Fault Trees

Fault trees are based on combinations of AND and OR gates. An example with an OR gate is shown in Figure 8-5. Similar to reliability block diagrams, probabilities of basic events are added or multiplied through the gates. Probabilities are added through OR gates and multiplied through AND gates. Circles represent basic events and rectangles are used for descriptions. Fault trees are excellent for modeling entire safety systems including field devices. The example is Figure 8-5 shows that if either the fire detector, fire panel, or fire pump fails to operate when needed, the system will suffer a failure. Fault trees are of limited value in modeling logic systems however, because similar to block diagrams, they do not account for time dependent factors. As with block diagrams, more complex formulas can be used to calculate the probabilities of the basic events. In addition, fault trees, as with all models, only account for known events. In other words, if you're not even aware about a particular failure, you subsequently can't include it in the fault tree.

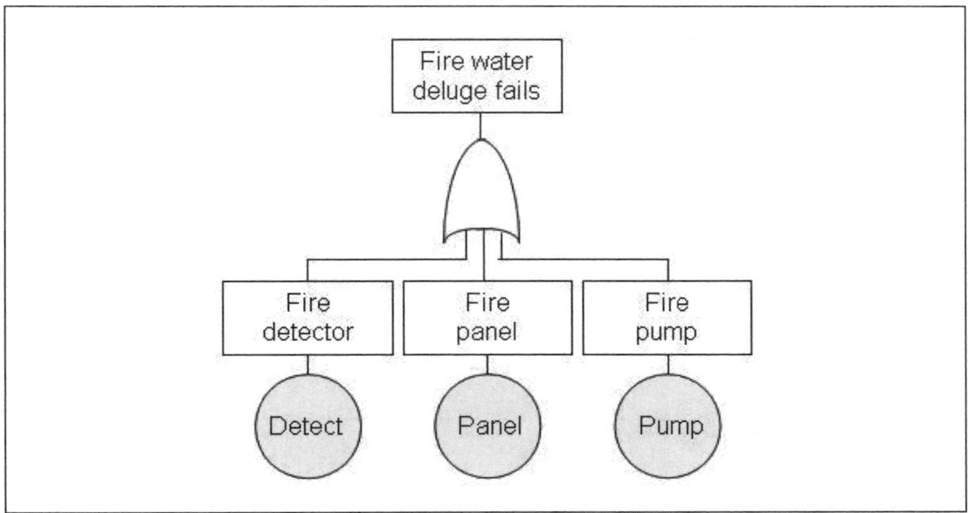

Figure 8-5: Fault Tree

8.7.3 Markov Models

Many reliability practitioners have settled upon Markov models. Markov modeling involves transition diagrams and matrix math. An example of a relatively simple model is shown in Figure 8-6. Markov modeling can be quite flexible. Unfortunately few are capable of using the technique, and those that can usually have advanced degrees.

Algebraic simplifications to Markov models have been available for decades (as developed in Reference 5 and listed References 11 and 12). These "simplified equations" are often associated with reliability block diagrams, but they can just as easily be incorporated into fault trees.

8.8 The Real Impact of Redundancy

Dual is not always better than single and triple is not always better than dual. Strange, but true. Refer to Figure 8-7 for the following discussion.

Let's start with a base case of a simplex (non-redundant) system. This can be referred to as 1oo1 (one-out-of-one). An example of a safe failure is

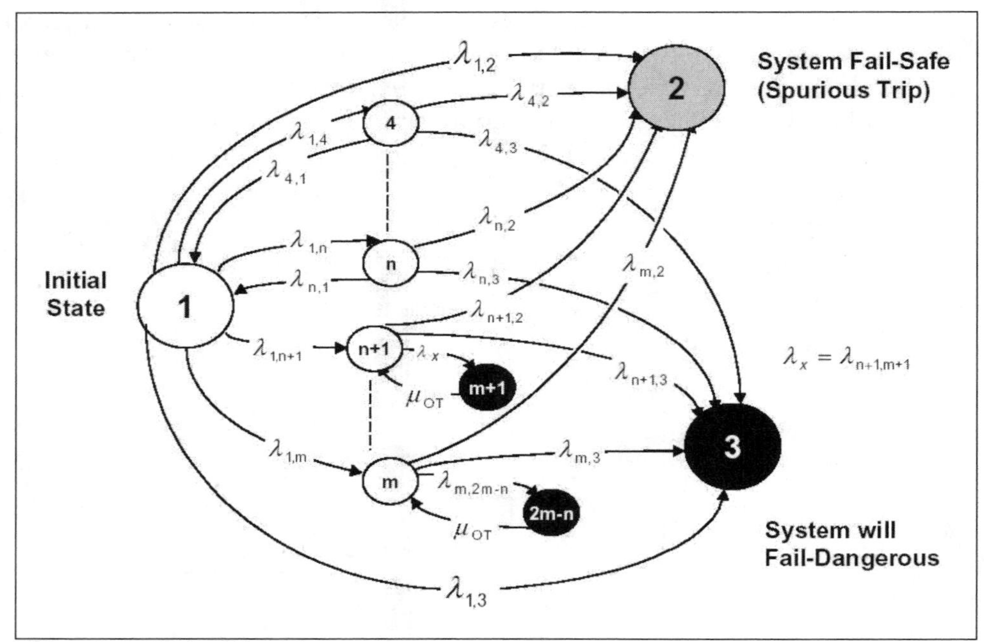

Figure 8-6: Markov Model

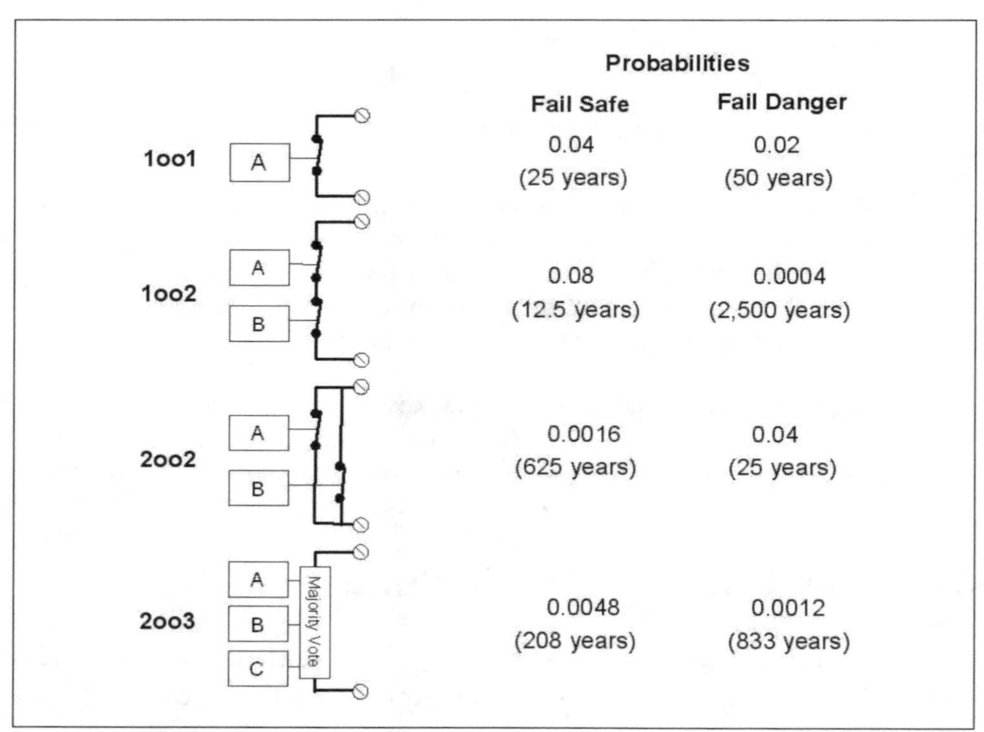

Figure 8-7: The Impact of Redundancy

where a relay contact opens and de-energizes the system causing a nuisance trip. Let's assume a failure probability in this mode of .04. This means that in a given time period (e.g., 1 year) the system has a 4% probability of suffering a nuisance trip. You could think of it as 4 systems out of 100 causing a nuisance trip within a year, or 1 system in 25 causing a nuisance trip, or a MTTF(safe) of 25 years. The numbers are just for comparison purposes at this point.

An example of a dangerous failure would be where the relay contacts are welded shut and won't operate when needed. Let's assume a failure probability in this mode of .02. This means that in a given time period (e.g., 1 year) the system has a 2% probability of not operating properly when there is a demand (i.e., a hazardous condition where the system must respond). You could think of it as 2 systems out of 100 not responding in a year, or 1 in 50 not responding in a year, or a MTTF(danger) of 50 years. You could make the safe and dangerous numbers the same, swap them around, or change their orders of magnitude, it really doesn't matter. Again, this is just to illustrate the impact of redundancy.

A dual 1oo2 (one-out-of-two) system has the outputs wired in series (assuming closed and energized contacts). One out of two means the system only needs one channel to perform a shutdown, hence the name one-out-of-two. If either channel can shut the system down, and there's twice as much hardware, there are twice as many nuisance trips. Therefore, the .04 doubles to .08. You could think of it as 8 systems out of 100 causing a nuisance trip within a year, or 1 system in 12.5 causing a nuisance trip, or a MTTF(safe) of 12.5 years.

In the dangerous mode, this system would fail to function only if both channels were to fail dangerously at the same time. If one were stuck, the other could still de-energize and shut down the system. What's the probability of two simultaneous failures? Actually, it's rather simple. What's the probability of one coin landing heads? 50%. What's the probability of two coins landing heads? 25%. This is just the probability of the single event squared (0.5 x 0.5 = 0.25). So the probability of two channels failing at the same time is remote (0.02 x 0.02 = 0.0004). You could think of it as 4 systems out of 10,000 not responding in a year, or 1 in 2,500 not responding in a year, or a MTTF(danger) of 2,500 years.

In other words, a 1oo2 system is very safe (the probability of a dangerous system failure is very small), but the system suffers twice as many nuisance trips as simplex, which is not desirable from a lost production standpoint.

A dual 2oo2 (two-out-of-two) system has the outputs wired in parallel. Here, both channels must de-energize in order to perform a shutdown. This system would fail to function if a single channel had a dangerous failure. Since this system has twice as much hardware as a simplex system, it has twice as many dangerous failures. Therefore the .02 doubles to .04. You could think of it as 4 systems out of 100 not responding in a year, or 1 in 25 not responding in a year, or a MTTF(danger) of 25 years.

For this system to have a nuisance trip, both channels would have to suffer safe failures at the same time. As before, the probability of two simultaneous failures is the probability of a single event squared. Therefore, nuisance trip failures in this system are unlikely (0.04 x 0.04 = 0.0016). You could think of it as 16 systems out of 10,000 causing a nuisance trip within a year, or 1 system in 625 causing a nuisance trip, or a MTTF(safe) of 625 years.

So a 2oo2 system protects against nuisance trips (i.e., the probability of safe failures is very small), but the system is *less safe* than simplex, which is not desirable from a safety standpoint. This is *not* to imply that 2oo2 systems are "bad" or should not be designed. If the PFD (which is the number we're concerned about from a safety standpoint) meets the overall safety requirements, then the design is acceptable.

Triple Modular Redundant (TMR) systems were developed back in the 1970s and released as commercial products in the early and mid 1980s. The reason for triplication back then was very simple: early computer-based systems had limited diagnostics. For example, if there were only two signals and they disagreed, it was not always possible to determine which one was correct. Adding a third channel solved the problem. One can assume that a channel in disagreement has an error, and it can simply be outvoted by the other two.

A 2oo3 (two-out-of-three) system is a majority voting system. Whatever two or more channels say, that's what the system does. What initially surprises people is that a 2oo3 system has a higher nuisance trip rate than a 2oo2 system, and a greater probability of a fail to function failure than a 1oo2 system. Some people initially say, "Wait a minute, that can't be!" Actually it is intuitively obvious, you just have to think about it a moment. How many simultaneous failures does a 1oo2 system need in order to have a dangerous failure? Two. How many simultaneous failures does a 2oo3 system need in order to have a dangerous failure? Two. Ah haa! A triplicated system has more hardware, hence three times as many dual failure combinations! (A+B, A+C, B+C) How many simultaneous failures does a 2oo2 system need in order to suffer a nuisance trip? Two. How many simultaneous failures does a 2oo3 system need in order to suffer a nuisance trip? Two. Same thing, a triplicated system has three times as many dual failure combinations. A triplicated system is actually a trade off. Overall, it's good in both modes, but not as good as the two different dual systems. However, a traditional dual system is either good in one mode or the other, not both.

If you look carefully at the numbers in Figure 8-7, you can see that the 1oo2 system is safer than 2oo3, and the 2oo2 system offers better nuisance trip performance than 2oo3. In theory, if a dual system could be designed to provide the best performance of both dual systems, such a system could actually outperform a triplicated system.

Improvements in hardware and software since the early 1980s mean that failures in dual redundant computer-based systems *can* now be diagnosed well enough to tell which of two channels is correct if they disagree. This means triplication is no longer required. The industry refers to this newer dual design as 1oo2D. [Ref. 15] Probably the most significant benefit for the dual 1oo2D vendors was getting their systems certified by independent agencies (e.g., TÜV and FM) to the same performance levels as the TMR systems. Unfortunately, the safety certifications do not cover nuisance trip performance. Therefore, the TMR vendors attack the dual vendors on this issue.

8.9 Basic Formulas

Algebraic simplifications to Markov models have been available for decades. The theory behind the following formulae is developed in Reference 5. Similar formulae are also shown in References 11 and 12. These are often called "simplified equations" and are usually associated with reliability block diagrams. This is actually incorrect as the formulae can just as easily be incorporated into fault trees.

The first set of formulae are for calculating mean time to failure, spurious ($MTTF^{spurious}$):

$MTTF^{spurious}$ formulae

1oo1: $1 / \lambda s$

1oo2: $1 / (2 * \lambda s)$

2oo2: $1 / (2 * (\lambda s)^2 * MTTR)$

2oo3: $1 / (6 * (\lambda s)^2 * MTTR)$

where:

MTTR = mean time to repair,

λ = failure rate (1 / MTBF)

s = safe failure

1oo1 stands for 1 out of 1, 2oo3 stands for 2 out of 3, etc.

The above formulae are valid when the repair rate is much greater than the failure rate (1/MTTR >> λ), or conversely when the MTBF is much greater than the MTTR. The formulae are based on the assumption that safe failures are revealed in all systems, even single channels of 2oo2 and 2oo3 configurations (e.g., through some form of discrepancy alarm). In other words, there are no safe undetected failures.

Average probability of failure on demand (PFD_{avg}) is calculated knowing the dangerous undetected failure rate and manual test interval. The PFD of dangerous *detected* failures can also be calculated (using slightly different formulae), but their impact on the final answer is insignificant, usually by more than an order of magnitude. The impact of dangerous detected failures can, therefore, be ignored in the calculations.

PFD_{avg} formulae for dangerous, undetected failures

1oo1: $\lambda_{du} * (TI / 2)$

1oo2: $((\lambda_{du})^2 * (TI)^2) / 3$

2oo2: $\lambda_{du} * TI$

2oo3: $(\lambda_{du})^2 * (TI)^2$

where:

TI = Manual test interval

λ_{du} = Dangerous undetected failure rate

The above formulae are valid when the MTBF ($1/\lambda_{du}$) is much greater than TI.

8.9.1 Impact Due to Manual Test Duration

If you test a safety system online (i.e., while the process is still running), a portion of the safety system must be placed in bypass in order to prevent shutting something down. The length of the manual test duration can have a significant impact on the overall performance of a safety system. During the test, a simplex (non-redundant) system must be taken offline. Its availability during the test period is zero. Redundant systems, however, do not have to be completely placed in bypass for testing. One leg, or slice, or a dual redundant system can be placed in bypass at a time. In effect, a dual system is reduced to simplex during a test, and a triplicated system is reduced to dual. The following formulae (for PFD_{avg}) were developed in reference 4 to account for online testing. The PFD from these formulae can simply added to the PFD formulae shown above.

1oo1: TD / TI

1oo2: $2 * TD * \lambda_d * (((TI / 2) + MTTR) / TI)$

2oo2: $2 * (TD / TI)$

2oo3: $6 * TD * \lambda_d * (((TI / 2) + MTTR) / TI)$

where:

TD = test duration

TI = test interval

8.10 Analysis of a Relay System

First, we must assume a failure rate (or MTBF) of a relay. Data books will show a considerable range of data for different types of relays. We can assume a 100 year MTBF for an industrial relay.

Next, we must consider how many relays to include in the calculation. Let's assume there will be one relay for each input and output in the system. Let's assume a relatively small interlock group, just eight inputs (e.g., high and low switches for pressure, temperature, level and flow) and two outputs (e.g., two valves). The system would suffer a nuisance trip if any of the ten relays were to fail open circuit. Therefore, we simply add the safe failure rate of ten relays. Remember that MTBF = $1/\lambda$. Assuming a relay is 98% fail-safe:

$MTTF^{spurious} = 1 / \lambda s$

$= 1 / ((1/100 \text{ years}) * 0.98 * 10)$

(0.98 represents the safe failure mode split,
10 represents the qty of relays)

= 10.2 years *(only use 2 significant figures: 10 years)*

We need to break the I/O down further for the PFD_{avg} calculation. When there is a shutdown demand placed on the system, it comes in on *one* input only. For example, only the high pressure shutdown goes into alarm, not all eight inputs at the same time. Also, the SIL is determined for each single function, so we should only model the PFD of a single function. Therefore, we should only include *one* input and both outputs in this particular model. This amounts to just three relays. Note that since a relay has no automatic diagnostics, all dangerous failures are dangerous undetected.

$PFD_{avg} = \lambda_{du} * (TI / 2)$

(1/100 years) * 0.02 * 3 * ((1 year/2))

(0.02 represents the dangerous failure mode split, 3 represents the qty of relays)

= 3.00 E-4

RRF = 1 / 3.00 E-4 = 3,300 (Risk Reduction Factor = 1 / PFD_{avg})

SA = 1–(3.00 E-4) = .9997 = 99.97% (Safety Availability = 1–PFD_{avg})

8.11 Analysis of a Non-redundant PLC System

What if we were to replace the relay system with a general purpose PLC? To be consistent with the I/O count of the relay examples, we can consider a PLC with 1 input and 1 output module. Do not attempt to split the calculation down to individual channels of a module, as there are usually common components on a board that could make the entire module fail. Let's assume:

CPU MTBF = 10 years

I/O module MTBF = 50 years

CPU safe failure mode split = 60%

I/O module safe failure mode split = 75%

We can also assume a dual redundant power supply. The probability of a simultaneous failure of two supplies would be orders of magnitude below the simplex components, so essentially it may be ignored in the model.

$MTTF^{spurious} = 1 / \lambda s$
$= 1 / (((1/10 \text{ years}) * 0.6) + ((1/50 \text{ years}) * 0.75 * 2))$
(0.6 and 0.75 represent the safe failure mode splits, 2 represents the qty of I/O modules)
$= 11$ years

Let's now assume 90% diagnostic coverage of the CPU and 50% for the I/O. The following assumes that the PLC is actually manually tested yearly.

$$PFD_{avg} = \lambda_{du} * (TI / 2)$$

$$[((1/10 \text{ years}) * 0.4 * 0.1) + ((1/50\text{years}) * 2 * 0.25 * 0.5)] * (1 \text{ year}/2)$$
(CPU failure rate, dangerous failure mode split, undetected portion) + (I/O module failure rate, quantity, failure mode split, undetected portion)

$$PFD_{avg} = 4.5 \text{ E-3}$$

$$RRF = 1 / (4.5 \text{ E-3}) = 220$$

$$SA = 1 - (4.5 \text{ E-3}) = .9955 = 99.55\%$$

So, the nuisance trip performance is about the same as the relays, but the PLC is *less safe* than relays by one order of magnitude! This tends to take a lot of people by surprise. Remember that we're assuming the PLC is actually fully tested manually once per year (which is an overly optimistic assumption for many systems). An I/O module diagnostic coverage of 50% is also an optimistic assumption for many general purpose systems. This is *not* saying anything negative about PLCs. General purpose PLCs are fine for what they were designed for—control. They simply were not designed for safety. Users are encouraged to get appropriate data from their vendors in order to model systems. Standards state that the vendors shall provide such information.

8.12 Analysis of a TMR System

The MTBFs and failure mode splits assumed for a PLC are reasonable to use even for a triple modular redundant – 2oo3 (TMR) system. The hardware is essentially the same, there's just more of it, and with greater diagnostic capabilities. As with the PLC, TMR systems have redundant power supplies which can initially be neglected in the model.

For a TMR system we can lump the input module, CPU, and output module together as one leg of a triplicated system. This would mean if input module #1, and CPU #2 both failed, the system would fail. Some TMR systems do in fact operate this way.

$MTTF^{spurious} = 1 / (6 * (\lambda_s)^2 * MTTR)$

$1 / (6 * (((1/10 \text{ years}) * 0.6) + ((1/50 \text{ years}) * 0.75 * 2))^2 * (4 \text{ hours} / 8760 \text{ hours per year}))$

(0.6 and 0.75 represent the safe failure mode splits, 2 represents the qty of I/O modules)

= 45,000 years

Let's initially assume 99% diagnostic coverage of both the CPU and I/O modules. The following assumes the TMR system is actually manually tested yearly, something that also may never actually happen with some systems.

$PFD_{avg} = (\lambda_{du})^2 * (TI)^2$

$[(((1/10 \text{ years}) * 0.4 * 0.01) + ((1/50\text{years}) * 2 * 0.25 * 0.01))^2 * (1 \text{ year} / 2)^2]$

(CPU failure rate, dangerous failure mode split, undetected portion) + (I/O module failure rate, quantity, failure mode split, undetected portion)

$PFD_{avg} = 6.25 \text{ E-8}$

RRF = 1 / (6.25 E-7) = 16,000,000

SA = 1 - (6.25 E-7) = .9999999375 = 99.99999375%

The calculations indicate that the TMR system would have an average time between nuisance trips of 45,000 years and a risk reduction factor exceeding ten million! Is there such a thing as SIL 7?! It's worth noting that TMR systems are not even certified for use in SIL 4. The IEC 61508 standard has a cautionary note essentially placing an upper limit on the level of performance that one can claim. Risk reduction numbers exceeding 100,000 don't really exist, especially for complex programmable systems. Could the calculations be wrong? Not really. There are simply factors we have not accounted for yet.

8.12.1 Common Cause

There are many redundant systems out in the world. Unfortunately, experience has shown that they do not perform as well as the calculations imply they should. The problem does not lie with the calculations, but rather the assumptions and what is (or is not) included in the calculations.

Everyone has heard the phrase, "Garbage in, garbage out." As with any modeling technique, what you don't include in the model won't show up in the results.

The assumptions in the TMR example above may be considered a bit simplistic. There was no accounting for common cause. Common cause can be defined as a single stressor that affects multiple components. As shown in Figure 8-8, A, B, and C represent a triplicated system. However, if item D fails, the entire system fails. Block D could represent a design error, programming error, maintenance error, EMI/RFI, or environmental problem, etc. One way of referring to common cause is called the Beta factor. This represents the percentage of failures identified in one slice or leg of the system that will affect the entire system. The Beta factor has been derived from empirical studies. Techniques for estimating Beta value ranges are described in References 5 and 12. Reference 5 lists a typical Beta value range of 20% when using redundant identical components (as is done with most systems).

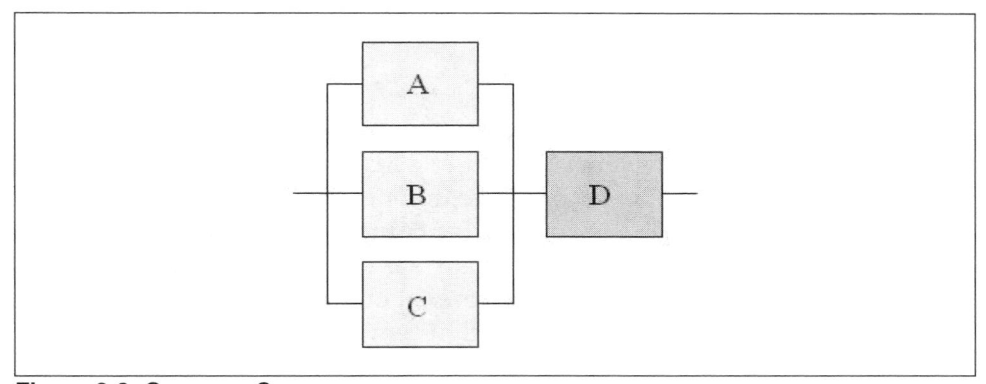

Figure 8-8: Common Cause

Common cause can easily be included in the model. One simply takes a certain percentage of the overall component failure rate in each failure mode of one slice of the redundant system and models it using the 1oo1 formulae. How much of an impact to the previous TMR answers would you expect when adding merely a 1% Beta factor to the calculations? Doing this for the $MTBF^{spurious}$:

$MTTF^{spurious} = 1 / \lambda s$

1 / ((((1/10 years) * 0.6) + ((1/50 years) * 0.75 * 2)) * 0.01)
(0.6 and 0.75 represent the safe failure mode splits, 2 represents the qty of I/O modules, 0.01 represents the 1% Beta factor)

= 1,100 years

In other words, our initial estimate of 45,000 years is overly optimistic. Just adding a 1% common cause Beta factor lowers the overall system performance *more than one order of magnitude.* Using a more realistic 10% Beta factor would lower the answer to 110 years. Common cause can obviously dominate overall system performance. The reader is encouraged to perform the PFD_{avg} calculation accounting for common cause.

Another way of accounting for common cause is referred to as systematic or functional failures. Rather than adding these failures as a percentage, a failure rate number is directly added to the 1oo1 portion of the formulae. While there are documented techniques for estimating a Beta value, it can be more difficult to justify a systematic failure rate.

8.13 Field Devices

All of the modeling examples above were for the logic boxes only. The performance requirements stated in the standards, however, are meant to include the field devices (i.e., sensors and final elements). What impact, if any, will this have?

Let's consider the same logic system as before, but this time with eight inputs (sensors) and two outputs (valves). Imagine a pressure vessel with an inlet and outlet line. The eight sensors could be switches to measure high and low values of pressure, temperature and level in the vessel, and flow to the vessel. The two valves could be inlet and outlet valves for the vessel. The logic could be that any sensor going into alarm would call for both valves to close. This represents eight safety instrumented functions (high pressure, low pressure, high temperature, low temperature, etc.). It doesn't matter that the valves are common to each function. Each function is designed to protect against a different hazard (e.g., the vessel bursting due to high pressure, the vessel drawing a vacuum and collapsing due to low pressure, a spoiled batch due to high temperature, etc.) Each hazardous event would have an associated frequency, consequence, and independent protection layers (if any). Therefore, each function would have its own SIL target (even though they may all be SIL 1).

Let's assume a sensor has a safe and dangerous MTBF of 50 years, and a valve has a safe and dangerous MTBF of 40 years. For the sensor, this would mean that out of 50 devices, in one year, one causes a nuisance trip, and when tested yearly, one is found not to be functioning properly. These are realistic numbers and are available from a wide variety of sources. Assuming any of the sensors could cause a nuisance trip (as described above), all eight should be included in the nuisance trip model. Either valve closing would stop production, so both valves should be included in the nuisance trip model. This brings the total number of field devices to 10 for this particular example.

The vast majority of installed field devices are non-redundant, even when used with TMR logic systems. To calculate the overall $MTTF^{spurious}$ of the system, we need to add the safe failure rates of all the components in the system (i.e., sensors, logic box, and final elements). The fault tree for this would look similar to Figure 8-5.

$MTTF^{spurious} = 1 / \lambda s$

1 / (((1 / 50 years) * 8) + (1 / 1,100 years) + ((1 / 40 years) * 2))
(8 sensors, TMR logic w/ Beta factor, 2 valves)

= 4.7 years

In other words, as far as nuisance trips are concerned, the TMR logic box is not the problem, the field devices are. What about the safety performance?

Many field devices are "dumb," meaning they have no form of self-diagnostics. For example, pressure switches and standard solenoid valves do not give any indication when they're stuck. The fault tree for this would again look similar to Figure 8-5. However, as described above, the vessel actually contains eight safety functions. The SIL is determined for each function. Therefore, we should model just *one* function, which in this example includes one sensor, the TMR logic box, and both valves. Assuming yearly testing, the PFD for the system can be determined by simply adding the PFDs of the three components.

Sensor PFD_{avg} = (1 / 50 years) * (1 year / 2) = 0.01

Logic PFD_{avg} = 6.25 E-8 (calculated earlier and not accounting for common cause)

Sensor PFD_{avg} = (1 / 40 years) * 2 * (1 year / 2) = 0.025

PFD_{avg} = 3.50 E-2

RRF = 1 / 3.50 E-2 = 29

SA = 1 - (3.50 E-2) = .965 = 96.5%

OUCH! In other words, our TMR logic box, which by itself could meet SIL 3 requirements, when combined with just a few non-redundant field devices, only meets SIL 1 (a risk reduction factor between 10 and 100)!

Again, the TMR logic box is not the source of the problem, the field devices are. A logic box does not a system make.

Some might conclude that TMR logic systems are not required, that one must focus on the field devices. This would not be entirely correct. Don't throw out the baby with the bath water. It only means that a chain is only as strong as its weakest link. In order to meet the requirements for the higher safety integrity levels, one requires either a) field devices with self-diagnostics, b) redundant field devices, c) devices with lower failure rates, d) more frequent testing, etc. All of these factors can be modeled using the formulae above. Section 8.13.1 covers one such example.

One might also conclude that one does not need a dual or triplicated logic systems for SIL 1 or 2 applications. This also would also not be entirely correct. Reliability predictions are a useful tool, but they should not be the only deciding factor. Not everything can be quantified. If you were in the market for a new car, would you buy one sight unseen just looking at a one-page specification sheet without even so much as a picture or test drive? There are intangible factors that can't be expressed numerically. Similarly, there are benefits with certain safety systems that make them ideally suited for safety, even in SIL 1 applications.

8.13.1 Partial Stroking of Valves

As covered in the prior section, field devices, especially valves, often represent the weakest link in a system. Traditional choices to improve performance include redundancy and/or more frequent manual testing. With valves, both choices are often financially and operationally unattractive. Redundant valves, especially large ones, can dramatically increase overall costs. The space required for extra valves may also not be present in an existing facility. Frequent full stroke testing of valves (e.g., quarterly) is usually not possible in continuous processes that are intended to run for years between maintenance turnarounds. One solution gaining in popularity since the late 1990s is partial stroke testing of valves. In fact, many vendors now offer packaged solutions.

The formulas used above can be modified to account for partial stroke testing of valves. Let's first start with the case of a single valve with a dangerous MTTF of 40 years and yearly full stroke testing.

$$PFD_{avg} = \lambda_{du} * TI/2$$

$$PFD_{avg} = (1 / 40 \text{ years}) * (1 \text{ year} / 2) = 1.25 \text{ E-2}$$

$$RRF = 1/PFD = 80$$

The above formula assumes the valve has no automatic diagnostics (i.e., all dangerous failures are undetected) and that the proof test is 100% effective (i.e., testing will reveal all failures). The resulting numbers fall into the SIL 1 range (although they are for a valve only and not an entire system). Increasing the test interval (i.e., less frequent testing) only degrades the numbers further.

One failure mode of a normally open valve is failing stuck open. Partial stroke testing would detect such a failure without impacting operations (i.e., not closing a valve and stopping production). The PFD formula can be modified to account for more frequent, although imperfect, partial stroke testing.

$$PFD_{avg} = (C * \lambda_d * TI_1 / 2) + ((1 - C) * \lambda_d * TI_2 / 2)$$

C = The diagnostic coverage factor

T_1 = The partial stroke test interval

T_2 = The full stroke test interval.

C, the diagnostic coverage factor, represents the percentage of failures that partial stroking would be able to reveal. Considerable controversy surrounds this figure. Numbers ranging from 60 to 90% have been claimed. If one were to choose 80% referencing the Pareto Principle (also known as the 80/20 rule), along with monthly partial stroke testing and yearly full stroke testing:

$$PFD_{avg} = (0.8 * (1/40 \text{ years}) * 0.083 \text{ year}/2) + (0.2 * (1/40 \text{ years}) * 1 \text{ year}/2)$$

$$PFD_{avg} = 8.3\text{ E-4} + 2.5\text{ E-3} = 3.3\text{ E-3}$$

$$RRF = 1/PFD = 300$$

The above numbers now fall into the SIL 2 range (RRF = 100 - 1,000). However, increasing the full stroke test interval to five years results in a RRF of 75, which is in the SIL 1 range (RRF = 10 - 100).

8.14 Fault Tolerance Requirements

A corollary of Murphy's law might be stated as, "If something can be abused, it will be." Reliability models are like captured foreign spies—if you torture them long enough they'll tell you anything. There have been cases where people claimed high safety integrity levels using non-redundant systems. If failure rates are selected low enough and test intervals are frequent enough, SIL 3 numbers can be achieved with a non-redundant system. However, numbers can lie.

Members of the IEC 61508 and 61511 committees felt it would be appropriate to include fault tolerance tables in their standards. This would hopefully limit people from abusing the math and modeling. The tables shown below are from the IEC 61511 standard (intended for end users in the process industry). The level of debate, arguments, and rewrites of these tables was significant. Two terms must first be defined in order to understand the tables.

Hardware fault tolerance refers to a level of required redundancy. For example, a hardware fault tolerance of 1 means that there are at least two devices, and the architecture is such that the dangerous failure of one of the two components or subsystems does not prevent the safety action from occurring. This would mean a 1oo2 configuration, *not* 2oo2. A fault tolerance of 2 requires a 1oo3 configuration.

The safe failure fraction (SFF) is one way of indicating the dangerous undetected failures of a device. Referring back to Figure 8-1 will help here. The safe failure fraction is defined as the number of safe failures, plus the dangerous detected failures, divided by the total. This can be expressed mathematically as:

$$SFF = (\lambda_{SD} + \lambda_{SU} + \lambda_{DD}) / l_{TOTAL}$$

This simply means that the greater the level of diagnostics, the less redundancy is required to meet any given SIL.

Table 8-2: Minimum Hardware Fault Tolerance for Programmable Logic Solvers

SIL	Minimum Hardware Fault Tolerance		
	SFF < 60%	SFF 60% to 90%	SFF > 90%
1	1	0	0
2	2	1	0
3	3	2	1
4	See IEC 61508		

Table 8-2 is for programmable (software-based) logic solvers. Table 8-3 is for non-programmable logic solvers and field devices. Considering the diversity of available field devices, some with no diagnostics, and some with very extensive diagnostics, the limitations associated with Table 8-3 should be evident. To make matters even more confusing, there are conditions where the fault tolerance requirements shown in Table 8-3 should *increase* by one (e.g., if the dominant failure mode is not to the safe state or dangerous failures are not detected), and other conditions where it may be *decreased* by one (e.g., prior use issues). Either way, Table 8-3 clearly indicates that in order to meet the higher SILs, redundant field devices are needed. If one has more detailed data on SFF of field devices, one can use more detailed tables found in IEC 61508.

Table 8-3: Minimum Hardware Fault Tolerance for Field Devices and Non-Programmable Logic Solvers

SIL	Minimum Hardware Fault Tolerance
1	0
2	1
3	2
4	See IEC 61508

8.15 Sample SIS Design Cookbook

Using the formulas above and valid failure rate data for the devices being considered, one could develop an SIS design "cookbook." An example is shown in Table 8-4. Like any cookbook, there are limitations on its use. For example, a recipe that makes a cake for four people cannot necessarily be scaled up to serve a wedding party of four hundred. As long as your design falls within the assumptions used in the development of your cookbook (e.g., quantity of devices, failure rates, test intervals, etc.), then each system need not be quantitatively analyzed in detail.

Table 8-4: Sample SIS Design Cookbook

SIL	Sensors	Logic	Final Elements
1	Switches or transmitters	Relays, Safety Solid State, General-Purpose PLCs, Safety PLCs	Standard valves
2	Transmitters with comparison[1], Safety transmitters[2]	Relays, Safety Solid State, Safety PLCs	Redundant valves, Simplex valves with partial stroking[3]
3	Redundant transmitters	Relays, Safety Solid State, Redundant Safety PLCs	Redundant valves with partial stroking
4	See IEC 61508		
Assumptions: 1 sensor, 2 valves, yearly testing			

Notes:
[1]Transmitters with comparison means using a single safety transmitter and a single BPCS transmitter and comparing the two signals. This would increase the level of diagnostic coverage significantly.
[2]Safety transmitters are available from several vendors and offer diagnostic coverage factors exceeding 95%.
[3]Partial stroke testing means testing a valve partially online without fully stroking it. There are a number of manual and automated methods of accomplishing this. The partial stroke needs to be carried out at least monthly to have a significant benefit.

8.16 Engineering Tools Available for Analyzing System Performance

It's useful to first model systems by hand so one realizes a) there is no magic involved, and b) you do not need a Ph.D. in math to do it. Crunching numbers by hand, however, can quickly become tedious and boring. Thankfully, there are engineering tools available to automate the modeling process.

One could simply take the formulas presented earlier and incorporate them into a computer spreadsheet. Along with calculating system performance, one can generate charts and diagrams showing the impact of test intervals, diagnostics, etc. This form of automation greatly aids in understanding which factors have a significant impact on system performance, and which don't.

There are a number of commercial programs for performing reliability block diagrams, fault trees, and Markov modeling. [Ref. 13] These programs are generic, however, and are not specific to the unique nature and design of most of the safety systems currently in use.

Control and safety system vendors have developed their own modeling programs. These programs may or may not be made available for users. There are other commercial programs specifically designed to model the performance of control and safety systems. [Ref. 14, 16, and 17]

Summary

Things are *not* as intuitively obvious as they may seem. Dual is *not* always better than simplex, and triple is *not* always better than dual. Which technology to use, what level of redundancy, what manual test interval, what about the field devices? If answers to these questions were easy, it would not have taken the standards groups 10 years to write their documents, and there would be no need for this book.

We do not design nuclear power plants or aircraft by gut feel or intuition. As engineers, we must rely on quantitative evaluations as the basis for our judgments. Quantitative analyses may be imprecise and imperfect, but it nevertheless is a valuable exercise for the following reasons:

1. It provides an early indication of a system's potential to meet the design requirements.
2. It enables one to determine the weak link in the system (and fix it, if necessary).

In order to predict the performance of a system, one needs performance data of all the components. Information is available from user records, vendor records, military style predictions, and commercially available data bases from different industries.

When modeling the performance of an SIS, one needs to consider two failure modes. Safe failures result in nuisance trips and lost production. The preferred term in this mode for system performance is mean time to failure, spurious ($MTTF^{spurious}$) which is usually expressed in years. Dangerous failures are where the system will not respond when required. Common terms used to quantify performance in this mode are probability of failure on demand (PFD), risk reduction factor (RRF), and safety availability (SA).

There are a number of modeling methods used to predict safety system performance. The ISA technical report ISA-TR84.00.02-2002, Parts 1-5, provide an overview of using simplified equations (to Markov models), fault trees, and full Markov models. Each method has its pros and cons. No method is more right or wrong than any other. All involve simplifications and account for different factors. (It's worth noting that modeling the same system in ISA-TR84.00.02, using all three methods, produced the same answers.) Using such techniques, one can model different technologies, levels of redundancies, test intervals, and field device configurations. One can model systems using a hand calculator, or develop spreadsheets or stand-alone programs to automate and simplify the task.

References

1. Kletz, Trevor A. *Computer Control and Human Error*. Gulf Publishing, 1995.
2. Leveson, Nancy G. *Safeware - System Safety and Computers*. Addison-Wesley, 1995.
3. Neumann, Peter G. *Computer Related Risks*. Addison-Wesley, 1995.

4. Duke, Geoff R. "Calculation of Optimum Proof Test Intervals for Maximum Availability." *Quality & Reliability Engineering International* (Volume 2), 1986. pp.153-158.

5. Smith, David J. *Reliability, Maintainability, and Risk: Practical Methods for Engineers.* 4th edition. Butterworth-Heinemann, 1993. (**Note:** 5th [1997] and 6th [2001] editions of this book are also available.)

6. ISA-TR84.00.02-2002, Parts 1-5. *Safety Instrumented Functions (SIF) – Safety Integrity Level (SIL) Evaluation Techniques.*

7. *OREDA-92. Offshore Reliability Data.* DNV Technica, 1992. (**Note:** Versions have also been released in 1997 and 2002.)

8. *Guidelines For Process Equipment Reliability Data, with Data Tables.* American Institute of Chemical Engineers - Center for Chemical Process Safety, 1989.

9. IEEE 500-1984. *Equipment Reliability Data for Nuclear-Power Generating Stations.*

10. *Safety Equipment Reliability Handbook.* exida.com, 2003.

11. ISA-TR84.00.02-2002, Parts 1-5. *Safety Instrumented Functions (SIF) – Safety Integrity Level (SIL) Evaluation Techniques.*

12. IEC 61508-1998. *Functional Safety of Electrical/Electronic/Programmable Electronic Safety-Related Systems.*

13. ITEM Software (Irvine, CA), 1998. Web site: www.itemsoft.com (retrieved 6/28/2005 from source)

14. CaSSPack (Control and Safety System Modeling Package). L&M Engineering (Kingwood, TX). Web site: www.landmengineering.com (Retrieved 6/28/2005 from source)

15. Goble, W.M. *Control Systems Safety Evaluation and Reliability.* Second edition. ISA, 1998.

16. SILver™ (SIL Verification). exida.com (Sellersville, PA) Web site: www.exida.com (Retrieved 6/28/2005 from source)

17. SIL Solver™. SIS-TECH (Houston, TX) Web site: www.sis-tech.com (Retrieved 6/28/2005 from source)

9

ISSUES RELATING TO FIELD DEVICES

Chapter Highlights

"When purchasing real estate, the three most important selection criteria are location, location, and location. When purchasing a Safety Instrumented System, the three most important selection criteria are diagnostics, diagnostics, and diagnostics."

9.1 Introduction

Field devices include sensors, sensing lines, final control elements, field wiring, and other devices connected to the input/output terminals of the logic system. These devices are often the most critical and, probably, the

most misunderstood and misapplied elements of safety systems. The emphasis paid toward field devices in the design and application of safety systems is disproportionally low compared to the potential impact that these devices can have on the overall system performance. It's estimated that approximately 90% of safety system problems can be attributed to field devices (see the analysis in Section 9.2).

Relays, solid state, and general-purpose programmable logic controllers (PLCs) have been used as safety system logic solvers. PLC's have the advantage of being software programmable, but they have very different dangerous failure mode splits compared to relays; therefore, their safety performance is very different. (These were the subjects of Chapters 7 and 8.) Many people (and naturally vendors) tended to focus on the difference between logic solver technologies. Since general-purpose and "safety" PLCs are now available with lower undetected dangerous failure rates and improved reliability, the focus is turning toward the field devices.

There are many publications available describing different types of field devices for various applications. The main objective of this chapter is to cover issues that need to be addressed when applying field devices in safety applications.

Some of these issues are as follows:

- diagnostics
- transmitters vs. switches
- smart transmitters
- smart valves
- redundancy
- inferential measurements
- specific application requirements

9.2 Importance of Field Devices

More hardware faults occur with the peripheral equipment – the sensors and final elements – of basic process control systems (BPCS) than with the control room equipment itself. The same also applies with safety system components.

9.2.1 Impact of Field Devices on System Performance

To demonstrate the effect of field devices on the performance of safety systems, consider the following example:

A safety instrumented system consisting of a pressure sensor, relay logic, a solenoid, and a shutdown valve is proposed for a SIL 1 application.

Based on the above SIL requirement, the PFDavg needs to be between 0.1 and 0.01.

The test interval is initially chosen to be once per year.

The PFDavg for each non-redundant element of the system can be calculated using the formula:

$PFD_{avg} = \lambda_d * TI/2$

where

λ_d = Dangerous failure rate of component

TI = Proof test interval of component

λ = 1 / MTTF (mean time to failure)

Equipment reliability and performance data for the proposed SIS is shown in Table 9-1.

Table 9-1: Reliability and Performance Data

Equipment Item	Mean Time To Failure, dangerous (years)	PFD_{avg}	PFD_{avg} % Contribution
Sensor	20	0.025	42
Logic system (4 relays)	100	0.005	8
Solenoid valve	33	0.015	25
Shutdown valve	33	0.015	25
Total	8.3	0.06	100

The overall risk reduction factor (RRF, which is 1/PFD) of the system is 16. The impact of the field devices on the overall system performance is 92%.

9.2.2 Percentage Split of System Failures

Figure 9-1 shows the percentage split of failures between the major elements of the system analyzed and summarized in Table 9-1.

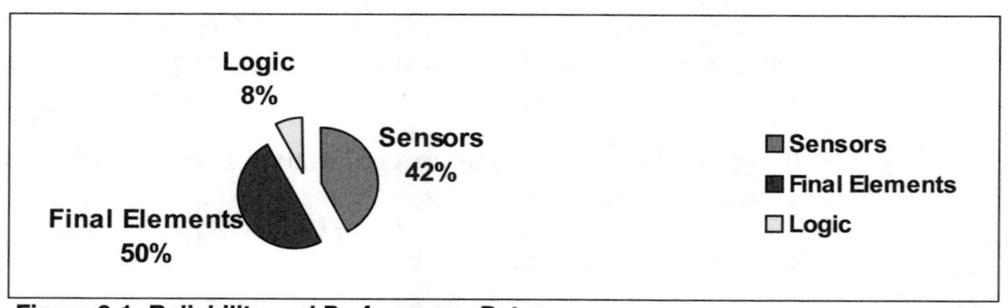

Figure 9-1: Reliability and Performance Data

As a general rule, field devices may account for approximately 90% of system failures. The logic system may account for only 10%. This is somewhat of a simplification as the numbers can vary significantly for different technologies and configurations.

The above data accounts for random failures only. Systematic failures (e.g., inadequate specifications, poor maintenance procedures, calibration errors, training, etc.) can also have a significant impact on the overall system performance. Systematic failures of field devices may be higher than for logic equipment because more activities may be centered on these devices.

9.3 Sensors

General issues relating to sensor technologies, failure modes, and diagnostics are discussed below.

9.3.1 General

Sensors are used to measure temperature, pressure, flow, level, etc. They may consist of simple pneumatic or electric switches that change state when a setpoint is reached, or they may contain pneumatic or electronic analog transmitters that give a variable output in relation to the process variable strength or level.

Sensors, like any other device, may fail in a number of different ways. They may fail safe and cause a nuisance or spurious trip (i.e., they may activate without any corresponding change of input signal). They may also fail dangerously (i.e., fail to respond to an actual demand or change of input signal). These are the two failure modes categories of most concern for safety systems. Other examples of sensor failures are:

- blocked impulse line
- leaking impulse line
- build-up of liquids in leg lines
- mechanical damage or internal leakage
- fouling of switch
- corrosion of contacts
- thermocouple burnout
- transmitter fails high or low
- smart transmitter left in "forced output" mode
- erratic output
- frozen transmitter signal
- drift

Most safety systems are designed to be fail-safe. This usually means that when the power source (electric, pneumatic or hydraulic) is lost, the safety system causes the process go to a safe state. This usually means shutting equipment down and halting production. Thought must be given as to how the sensors should respond in order to be fail-safe. Nuisance trips should be avoided for safety reasons as well since startup and shutdown modes of operation often involve the highest levels of risk. Also, a system that causes too many nuisance trips may be placed in bypass, resulting in a loss of all protection.

The general requirements for fail safe operation are:

- sensor contacts would be closed and energized during normal process operation
- transmitter signals would go to a trip state upon failure (or at least be detected by the system so corrective action could be taken)
- output contacts from the logic system would normally be closed and energized
- final elements would go to a safe position (e.g., valves would close) upon loss of air
- other final devices (e.g., motors or turbines) would stop

Transmitters can usually be configured to fail upscale or downscale in the event of an internal electronics failure. However, *not* all failures can be detected. Therefore, there is no guarantee that the device will always achieve the desired state upon failure. Thought should be given to the failure mode for each sensor. For example, it may be desirable to have a low level transmitter fail low, but a high pressure transmitter fail high. An overall, across-the-board recommendation for all sensors cannot be made. (As is common for control valves, the failure mode for transmitters should be defined and included in the safety requirement specification (SRS); see Chapter 5.)

Some measurements may be inferred from other variables. For example, if a system is designed to shut down due to high pressure, it may be effective to also monitor temperature (if, due to the process, an elevated temperature might also imply a high pressure). Inferential measurements can be used to initiate shutdowns. The following should be taken into consideration when using such measurements:

- **Response time**: The response of the measured variable needs to be within the process safety time as defined in the requirements specification.
- **Correlation**: The relationship between the two variables should be well understood.
- **Accuracy**: The relationship should provide an accurate representation.

If there are uncertainties with the inferred variable, consideration should be given to using the measurement for alarm purposes. The operator can then make the final decision based on a review of other process parameters. Inferred variables can also be used as extended diagnostics tools. Comparing a transmitter or switch output with its inferred value or status can be used to trigger diagnostic alarms.

Special care should be taken when operating sensors at the extreme end of their ranges due to potential low accuracy problems. For example, a sensor designed to operate at 1,000 psi may not be able to differentiate between 20 and 25 psi. Most primary elements have their accuracies stated in percentage of full scale. A transmitter with a stated accuracy of 1% may actually have an error of 100% while measuring the first 1% of its scale.

Sharing of sensors with the BPCS is generally not recommended for the following reasons:

- The SIS may need to be called into action if a component of the BPCS fails. The failure of the BPCS could be a sensor, logic system, or final element. If a sensor in the BPCS fails thereby creating a potential hazard, then trying to operate the SIS from the same sensor that's failed makes little sense.
- The procedures for testing, calibration and maintenance of the SIS are usually more stringent than for the BPCS. The integrity of the safety system may be compromised if changes are made to its sensors without following adequate procedures.
- Online testing of the shared field devices may not be possible.

The above rationale also applies to final elements.

9.3.2 Switches

Discrete switches do not provide any form of diagnostic information. For example, if under normal operation a pressure switch has a closed set of alarm contacts that are designed to open upon low pressure, and the contacts become stuck and can't open as a result, the sensor has no means to detect or annunciate the failure. The only way to know whether such a device is working is to periodically *test* it.

Discrete/on-off switches have been used in the past for the following reasons:

- most shutdown functions are discrete (e.g., on/off, open/close)
- discrete relay logic was widely used in the past

- PLCs were originally designed for discrete inputs and outputs (I/O)
- lower cost

In order to address some of the disadvantages of dormant, discrete devices (i.e., the lack of diagnostics), an on/off pneumatic indicating controller with a built-in sensor could be used as the primary sensor. The set point of the controller would be set at the process trip value. The output signal from the pneumatic controller would be either 3 or 15 psi. The signal could be tubed to a pressure switch set at 9 psi, with the contacts from this switch wired to the electric (relay), electronic (solid state), or software-based (PLC) logic system.

Some advantages of using on/off controllers are:

1. local indication of process variable
2. ease of partial online testing of the sensor – the set point of the controller could be changed to initiate a trip
3. improved accuracy, rangeability, and reliability as compared to switches
4. the controller can be key locked for security

9.3.3 Transmitters

Transmitters provide an analog signal in relation to the input variable. This indicates, at least in a limited sense, whether the device is functioning. Any information is better than none. However, if the transmitter dynamic output is never monitored for changes by the operators or the logic system, there may be no more usable information than with a discrete switch. It would be like having a color printer, but only printing black and white documents. The perceived benefit of having the color printer is illusory if one is unable to use it.

Even though they may cost more than discrete switches, analog transmitters are often preferred for the following reasons:

- increased diagnostics
- field indicator available
- lower safe and dangerous failure rates
- comparison of signal with BPCS
- single transmitter can replace several switches
- better accuracy and repeatability than switches

It is often possible to preset the failure mode of transmitters (i.e., 0% or 100% reading). (Note: If a sensor only has 30% diagnostics—which can be the limit for many—then the sensor will simply not always go to the selected failure mode when desired.) Also, by analyzing the 4–20 mA signal from the transmitter, it's possible to determine the operational status of the device (e.g., <4 mA or > 20 mA would indicate that the transmitter has suffered a failure).

One may also consider the use of highway addressable remote transducer (HART) alarm monitors. (Devices for other protocols are also available, but this discussion will be restricted to HART.) Smart transmitters with HART protocol are capable of providing a wealth of data beyond the simple 4–20 mA measurement signal. By connecting a HART alarm monitor to a 4–20 mA transmitter, the monitor can provide transmitter fault signals in addition to the process signal. This has the potential to improve the diagnostic coverage and improve the safety performance of the overall system. The monitor can initiate transmitter fault signals if any of the following conditions occur:

- transmitter malfunction
- sensor input out of range
- transmitter analog output frozen
- analog output out of range

By utilizing the above diagnostics, the diagnostic coverage of the transmitter may be increased approximately 20%. A detailed failure mode, effects and diagnostic analysis (FMEDA) would be required to determine the precise level of diagnostic coverage.

The two signals from the transmitter and alarm monitor could be wired to the logic system to generate either an alarm or shutdown signal (see Section 9.5 and Table 9-2 for further analysis).

9.3.4 Sensor Diagnostics

Improving the diagnostics (i.e., diagnostic coverage) of a safety system improves the safety performance of the system. In other words, it decreases the probability of failure on demand (PFD), or increases the risk reduction factor (RRF, which is 1/PFD). The following formula shows the impact of diagnostic coverage on the dangerous failure rate.

$\lambda^{DD} = \lambda_d C_d$
$\lambda^{DU} = \lambda_d (1\text{-}C_d)$

C_d = Diagnostic coverage factor of dangerous failures
λ_d = Total dangerous failure rate
λ^{DD} = Dangerous detected failure rate
λ^{DU} = Dangerous undetected failure rate

Increasing the diagnostic coverage factor of dangerous failures (C_d) increases the dangerous detected failure rate (λ^{DD}) and decreases the dangerous undetected failure rate (λ^{DU}). Note that diagnostics do not change the *overall* failure rate, just the percentage of failures that are detected.

One advantage of increasing diagnostics is that the safety system designers have the option to shut the process down or only generate an alarm if a dangerous fault is detected in a sensor. In other words, rather than stopping production due to a failed sensor, the system can bypass the failed signal until repairs are made. Procedures must naturally be in place in order to assure safety while a device is in bypass.

The ideal would be to have sensors with 100% diagnostics and, therefore, no dangerous undetected failures. Unfortunately, 100% diagnostics is not possible, yet the objective is to get it as high as possible.

A more recent parameter used to indicate the effectiveness of diagnostics is the safe failure fraction (SFF). The safe failure fraction is defined as the ratio of the safe failure rate plus the dangerous detected failure rate, divided by the total failure rate. This can be expressed mathematically as shown below. The safe failure fraction can be determined for each component in a system (i.e., sensor, logic, or final element).

$$SFF = \frac{\lambda^{SD} + \lambda^{SU} + \lambda^{DD}}{\lambda^{SD} + \lambda^{SU} + \lambda^{DD} + \lambda^{DU}} = 1 - \frac{\lambda^{DU}}{\lambda^{SD} + \lambda^{SU} + \lambda^{DD} + \lambda^{DU}}$$

The need for periodic maintenance and manual testing can be reduced by improving sensor diagnostics.

One simple, yet effective method for improving sensor diagnostics is to compare the signal from a transmitter with other related variables (e.g., the control system transmitter reading the same process variable). A discrepancy between the signals would indicate problems that a single device might not be able to diagnose on its own. Also, smart transmitters generally provide a higher level of diagnostics than conventional transmitters (see Section 9.3.5 for more details).

Transmitters are now available that are certified for safety applications. These are frequently called "safety transmitters" to distinguish them from standard transmitters. The primary difference is that these devices provide a higher level of diagnostics, and often redundant internal circuits, than conventional or smart transmitters. Rainer Faller's *Safety-Related Field Instruments for the Process Industries* provides additional information on the use and application of these transmitters. [Ref. 4]

9.3.5 Smart Transmitters

Some of the advantages of smart transmitters (i.e., transmitters with additional diagnostics and digital communications, rather than a "conventional" analog transmitter with only a 4–20 mA output signal) in safety applications are:

1. better accuracy compared to conventional transmitters
2. negligible long term drift and excellent stability
3. improved diagnostics: diagnostic coverage factors and SFF greater than 80% are now being obtained based on failure mode, effects, and diagnostic analysis (FMEDA)
4. greater predictability of failure modes
5. remote maintenance calibration capabilities *
6. less exposure by maintenance technicians to hazardous environments *
7. special safe work permits may not be required because of remote maintenance capabilities *

* Note that the adjustment of process-related parameters of sensors needs to be protected and controlled against unauthorized and/or inadvertent changes.

Some *dis*advantages of smart transmitter are:

1. ease of changing parameters (e.g., range) without ensuring that changes are reflected in the logic and other systems
2. ease of leaving a transmitter in test or "forced output" mode
3. cost – smart transmitter are more expensive than conventional transmitters
4. response may be slower than conventional analog transmitters

Even with the above disadvantages, smart transmitters are increasingly being used in safety applications.

9.4 Final Elements

General issues relating to valves and diagnostics are discussed below.

9.4.1 General

Final elements are the devices used to implement a safe shutdown of a process. The most common final elements are solenoid operated valves that provide air to a diaphragm or piston actuator. Removal of the air pressure usually results in the valve going to its safe state.

Final elements generally have the highest failure rates of any component in the overall system. They're mechanical devices subject to harsh process conditions. Shutdown valves are usually open and not activated for long periods of time, except for testing. One common dangerous failure of such valves is being stuck. One common safe failure is the solenoid coil burning out.

Valves should be fail safe upon loss of power. Diaphragm and piston actuated valves normally require spring-loaded actuators or pressurized volume bottles in order to be fail safe. Double-acting valves (i.e., air to open and air to close) are used when valves are required to "hold" last position upon air failure, or when the size and cost associated with spring-loaded single-acting actuators is not feasible.

Typical failure modes for the final elements are:

Solenoid Valves

- coil burnout
- plugging of ports or vents
- corrosion of terminals or valve body making solenoid inoperative
- hostile environment and/or poor quality instrument air resulting in sticking

Shutdown Valves

- leaking valve
- actuator sizing insufficient to close against shut-off pressure
- sticky valve stem or valve seat

- blocked or crushed air line
- stuck

Some use the BPCS control valve also as a shutdown valve. Some perceive a benefit because the control valve is normally moving, therefore they consider it to be "self testing." As with sensors, sharing of final elements with the BPCS is generally not recommended (see Section 9.3). Additional reasons why process control valves also should not be used as shutdown valves are as follows:

- Control valves account for the largest percentage of instrument failures.
- Process control valves are designed for their control capabilities. There are other parameters of importance for shutdown valves that may not be considered when selecting control valves (e.g., leakage, fire safety, speed of response, etc.).

In spite of these generalizations, process control valves may be used to provide shutdown functions in cases where it may be impractical or not cost effective to install a separate shutdown valve. Recommendations in such cases might include tubing a solenoid valve directly to the actuator bypassing the valve positioner, or the use of redundant solenoids. Careful sizing of the solenoid is also required in order to ensure adequate speed of response of the main valve. A safety case should be developed to show that the risks associated with using a single shared valve are acceptable.

Some potential exceptions to the above generalizations are motor starters and non-electric drive systems.

9.4.2 Valve Diagnostics

Shutdown valves must operate when a signal is sent from the logic system to the valve. During normal operation these valves are usually fully open and will remain in this state for long periods of time, unless tested or activated in some way by the logic system. Valve diagnostics should be considered both when the valve is in its normal steady state (e.g., open), as well as when the valve is actuated.

1. **Normal operation**: During normal operation the following techniques can be used to diagnose the state of the valve:
   - Online manual testing (refer to Chapter 12)
   - Using a smart valve positioner (refer to Section 9.4.3)
   - Alarm if the valve changes state without a signal from the logic system

- Partial stroke testing. Testing the valve automatically by allowing the logic system to automatically move the valve a small percentage and monitoring movement. An alternative could be to have the logic system send a signal to the valve for a fixed time (usually a few seconds, based on valve and process dynamics). If the logic system fails to see a movement by means of a position indicator within a specified period of time, an alarm is activated indicating a potential problem with the system. Additional components may need to be installed in order for this to be accomplished. There are a variety of other techniques for accomplishing partial stroke testing.

2. **When activated**: Limit switches or position indicators installed to provide feedback indicating that the valve has operated correctly or incorrectly. Such diagnostics should *not* be considered "active". If the valve has not stroked for a long period of time, limit switches will not indicate when the valve actually fails stuck.

9.4.3 Smart Valve Positioners

Some have installed smart valve positioners on shutdown valves in order to obtain additional diagnostic data on the health of the valves. The positioners can be installed simply to provide diagnostic data, or they may be used to trip the valve. In the latter case, the shutdown signal from the logic box is connected directly to the positioner. Information that can be obtained from smart valve positioners includes:

- position feedback
- problems with the air supply or actuator
- limit switch status
- valve position status
- alarm and self-diagnosis of positioner

Many smart positioners provide the capability for partial stroke testing of the shutdown valve either locally or remotely. Partial stroke testing is one way of meeting the requirements for SIL 2 and SIL 3 applications. The partial stroke test usually moves the valve about 10% and provides diagnostic coverage factors ranging between 60 to 90%. The valve would be completely tested (i.e., closed) during the major plant turnaround. There are a variety of commercially available (and certified) partial stroking solutions. Section 8.13.1 shows the performance impact of partial stroke testing.

9.5 Redundancy

If the failure of any single sensor can't be tolerated (i.e., a nuisance trip or a fail-to-function failure), or if a single sensor can't meet the performance requirements, then redundant or multiple sensors may be used. Ideally, the possibility of two sensors failing at the same time should be very remote. Unfortunately, this doesn't account for common cause or systematic failures that might impact multiple sensors at the same time. Common cause failures are usually associated with external environmental factors such as heat, vibration, corrosion and plugging, or human error. If multiple sensors are used, they should be connected to the process using different taps (where possible) as well as have separate power supplies so as to avoid common cause plugging and power failures. Consideration may also be given to using sensors from different manufacturers, as well as having different maintenance personnel work on the sensors, so as to avoid the possibility of design or human errors.

Whenever redundant transmitters are used, consideration should be given to utilizing different technologies (e.g., level sensing using both a differential pressure measurement and a capacitance probe). Both sensors should have a proven performance history in such applications. If only one sensor has a known proven performance record, then it may be best to use only that device for both sensors. While the use of diverse technology results in minimizing the potential of common cause problems, it does result in other potential drawbacks that should be considered (e.g., extra stores, training, maintenance, design, etc.).

Diverse redundancy utilizes different technology, design, manufacture, software, firmware, etc., in order to reduce the influence of common cause faults. Diverse redundancy should be considered if it's necessary to lower the common cause in order to achieve the required SIL (refer to Section 8.12.1). Diverse redundancy should not be used where its application may result in the use of lower reliability components that may not meet the reliability requirements for the system. Systematic failures are not prevented through the use of redundancy, although they may be reduced utilizing diverse redundancy.

9.5.1 Voting Schemes and Redundancy

The redundancy and voting that may be required depends on a number of different factors (e.g. SIL requirements, desired spurious trip rate, test frequency, component failure rates and modes, diagnostics, etc.). The following table shows the configurations commonly used with sensors and final elements. Section 8.8 provides a more detailed discussion on redundancy.

Sensors	
1oo1	1 out of 1 (i.e., a single sensor). Used if the component meets the performance requirements.
1oo1D	1 out of 1 with diagnostics. The diagnostics may be provided by an additional alarm monitor or built into the sensor (i.e., a certified safety sensor). See Table 9-2 for details.
1oo2	1 out of 2. Two sensors installed, but only *one* is required to shutdown. This configuration is *safer* than 1oo1, but has twice as many nuisance trips.
1oo2D	1 out of 2 with diagnostics. The diagnostics may be provided by an additional alarm monitor or built into the sensor. This configuration is "fault-tolerant" and can survive single safe or dangerous failures and continue to operate properly.
2oo2	2 out of 2. Two sensors installed and *both* are required to shutdown. This scheme is *less* safe than 1oo1, yet has fewer nuisance trips.
2oo3	2 out of 3. Three sensors installed and two are required to shutdown. As with 1oo2D, this configuration is also "fault-tolerant" and can survive single safe or dangerous failures and continue to operate properly.
Final Elements	
1oo1	Single valve. Used if the device meets the performance requirements.
1oo2	1 out of 2. Two valves installed, but only *one* is required to shutdown. As with sensors, this scheme is safer that 1oo1, but is susceptible to twice as many nuisance trips.
2oo2	2 out of 2. Two valves installed and *both* are required to shutdown. As with sensors, this scheme is *less* safe than 1oo1, yet has fewer nuisance trips.

9.5.1.1 Voting Logic for 1oo1D Configurations

Figure 9-2 shows how a single smart transmitter with diagnostics can be used to reduce the dangerous undetected failure rate λ^{DU}.

For the above system, the transmitter would incorporate diagnostics and the HART protocol. The HART alarm monitor checks the transmitter's HART digital signal and provides an alarm if either the process variable is outside set limits or the diagnostics within the HART monitor detect a failure. This is similar to the operation of safety transmitters. Table 9-2 below shows the possible conditions for a single transmitter.

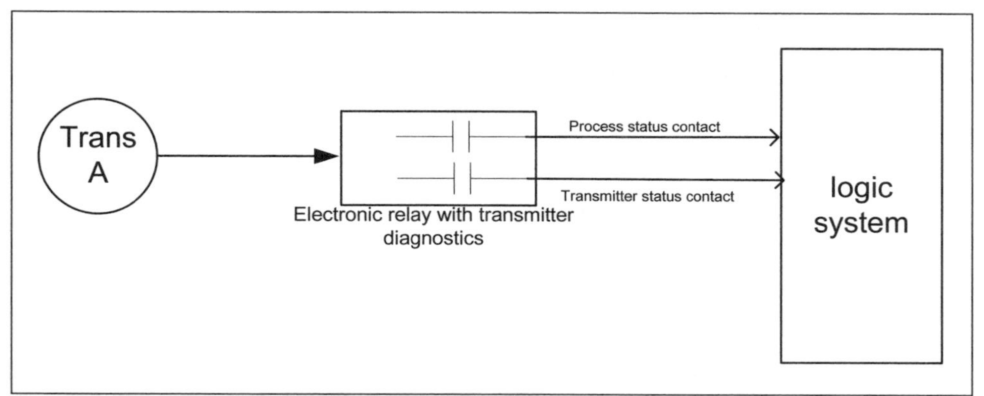

Figure 9-2: 1oo1 Transmitter with Diagnostics

Table 9-2: 1oo1D Logic States

Condition	Process state as sensed by transmitter 1 = No Trip 0 = Trip State	Transmitter state sensed by diagnostics 1 = No Error 0 = Fault Detected	1oo1D logic status
1	1	1	OK
2	0	1	Trip
3	1	0	Trip or Alarm
4	0	0	Trip or Alarm
Note: An alarm should always be generated if a transmitter fault occurs.			

Condition1: The process state is normal (OK) and no faults have been detected in the transmitter.

Condition 2: The process state is outside normal limits and no faults have been detected in the transmitter. Trip state.

Conditions 3 & 4: The process state is assumed to be unknown since a fault has been detected with the transmitter. The logic system can be programmed to either trip or alarm under such conditions. This will impact the overall safe and dangerous performance of the system depending on which selection is made.

The PFD_{avg}, RRF (1/PFD), and $MTTF^{spurious}$ values for a typical transmitter in 1oo1 and 1oo1D configurations can be calculated. Assuming the following data:

$MTTF_S =$ 25 years

$MTTF_d = 50$ years

$TI = 1$ year

$C_d = 60\%$

$MTTR = 12$ hrs

$\lambda = 1/MTTF$

$\lambda^{DD} = \lambda_d C_d$

$\lambda^{DU} = \lambda_d(1\text{-}C_d)$

For 1oo1 architecture

$$PFD_{avg} = \lambda_d * TI/2$$

$$= 0.02 * 1/2$$

$$= 0.01$$

$$MTTF^{spurious} = 1/\lambda_s$$

$$= 25$$

For 1oo1D architecture

$$PFD_{avg} = \lambda^{DD} \cdot MTTR + \frac{\lambda^{DU} \cdot TI}{2} = 0.004$$

$$MTTF^{spurious} = 1/\lambda_s$$

$$= 25$$

	PFD_{avg}	RRF	$MTTF^{spurious}$ (years)
1oo1	0.01	100	25
1oo1D	0.004	250	25

There is a decrease, although not significant, in the PFD_{avg} for the 1oo1D architecture modeled above. A certified safety transmitter would offer even more improved performance as its diagnostic coverage factor would be in the range of 90%.

9.6 Design Requirements for Field Devices

Field devices are integral parts of the overall safety system. They must, therefore, be selected, designed and installed in order to satisfy the performance requirements (SIL) of each safety instrumented function (SIF).

As discussed in Chapter 3 and other sections above, field devices for safety systems should be separate and independent of the basic process control system (BPCS) and other non-safety-related systems. (There are always exceptions, such as if redundant sensors are used, so long as no single failure can cause a problem.) Sharing sensors with the BPCS may create maintenance and procedural problems (e.g., security, range change, testing, etc.). To achieve the necessary independence, the following components should be separate from the BPCS.

- Field sensors, taps, root valves, and impulse lines
- Final elements (e.g., solenoid valves, main shutdown valves, etc.)
- Cables, junction boxes, air lines and termination panels associated with the I/O, and dedicated panels for the safety system
- Power supplies

If safety system field device signals are to be sent to the BPCS for data logging or comparison, the signals between the two systems should be isolated (e.g., using optical isolators) in order to prevent a single failure from affecting both systems.

Field device installations need to conform to applicable local/state codes and any pertinent federal regulations. Field devices should be "fail safe." In most cases this means that the devices are normally energized. De-energizing or loss of power will initiate a shutdown.

Energize to trip systems are used to reduce potentially hazardous nuisance trips due to power failures, brownouts, component failures, and open wiring. In such cases, de-energized open circuits would use line monitoring, also called a supervised circuit, which requires an end of line device and monitoring of a trickle current. This allows the logic system to distinguish between a normally open circuit, closed circuit, wiring short, or wiring open circuit.

Only proven technologies should be used in safety applications. A proven and acceptable track record should be established before field devices are used in safety systems.

9.6.1 Sensor Design Requirements

The following general requirements should be considered:

- Fail safe systems should use normally closed contacts and normally energized circuits.
- Sensors should be connected directly to the logic system. They should not be connected to any other system that is not part of the shutdown system.
- Smart transmitters are being used more frequently for safety systems because of enhanced diagnostics and improved reliability. When using such transmitters, procedures should be established to ensure that they cannot be left in the "forced output" mode. Management of change and security access procedures need to be in place in order to allow configuration or calibration changes.
- Contacts for electrical switches should be hermetically sealed for greater reliability.
- Electronic trip amplifiers/current relays connected to dedicated safety system analog field transmitters may be used as inputs to a discrete logic system (e.g., relays).
- Field sensors associated with safety systems should be differentiated in some way from the BPCS sensors (e.g., unique tag, numbering system, or color).
- When redundant sensors are used, a discrepancy alarm should be provided to indicate the failure of a single sensor.

9.6.1.1 Flow Sensors

Although orifice plates represent the majority of applications and are "proven" and reliable enough in most applications, vortex and magnetic flowmeters may offer advantages such as ease of installation and improved performance.

9.6.1.2 Temperature Sensors

The primary failure mode of thermocouples is burnout; therefore, burnout detection and alarms should be used.

For non-fail-safe systems, upscale burnout indication for low temperature, and downscale for high temperature, may be considered.

9.6.1.3 Pressure Sensors

Pressure transmitters are easy to install and very reliable. The main precautions in selecting pressure devices are as follows:

- Range selection – ensure that the trip setting is within 30 to 70% of the transmitter range.
- Facilities for zeroing transmitters are required.
- Installation should ensure that condensate build up will not create calibration shift.
- The use of remote diaphragm seals instead of long leg lines can be effective in eliminating leg line problems.

9.6.1.4 Level Sensors

Bubbler systems with air or nitrogen purge have proven to be reliable and generally require little maintenance, although this naturally depends upon the application. Pressure transmitters with sealed capillary system are also very reliable and require very little maintenance. Floats, radar, sonic, capacitance probes, and nuclear devices are also used for many applications.

9.6.2 Final Element Design Requirements

Final elements should generally remain in their safe (shutdown) state after a trip until manually reset. They should be allowed to return to their normal state only if the trip initiators have returned to their normal operating conditions. The following should be considered in selecting final elements:

- opening/closing speed
- shutoff differential pressure in both directions of flow
- leakage
- fire resistance
- material suitability/compatibility
- diagnostic requirements
- fail-safe state of the valve (i.e., fail open or closed)
- need for position indicator or limit switch

- experience/track record of device
- capability for online maintenance

9.6.2.1 Actuators and Valves

The use of handjacks on shutdown valves should be avoided so the valves can't be left in a jacked or inoperable state. If handjacks are installed, they should be car-sealed and formal procedures should be in place to control their use.

Block and bypass valves may be considered for each valve that fails closed. Block valves alone may be provided for valves that fail open. For cases where both block and bypass valves are used, a bleeder valve may need to be installed upstream of the downstream block valve. Limit switches can be utilized to initiate an alarm if a bypass valve is opened or a block valve is closed. Some companies are satisfied controlling the operation of block and bypass valves through the use of car sealing and/or procedures.

In general, safety shutoff valves should be dedicated and separate from control valves (refer to Section 9.4 and Chapter 3). The sharing of valves for both control and safety is permitted, but the application must be carefully reviewed and justified.

9.6.2.2 Solenoids

Solenoid valves were originally designed to be installed in control room environments. With minor modifications for outdoor use, they are now used in the field where they are required to operate continuously under extreme conditions.

In general, the reliability of solenoids is low. As a result, they can be one of the most critical (weakest) components in the overall system. One common failure is the coil burning out which causes a false trip. It's important to use a good industrial grade solenoid, especially for outdoor use. The valve must be able to withstand high temperatures, including the heat generated by the coil itself, along with heat generated by furnaces, exposure to sun, etc. Dual coil arrangements are available to keep the solenoid energized if one coil were to burn out. Redundant solenoid arrangements can also be utilized.

Low power (24 VDC) solenoids appear to be more reliable because of the low power, hence low heat generation. Very low power solenoids, however, such as intrinsically safe (IS) types, may be extremely reliable in terms of coil burnout, but care needs to be taken in their selection as many are pilot operated. Some users have reported sticking and plugging problems with pilot-operated solenoids (although this is naturally application and environment dependent). Some users prefer the use of "direct connected" (i.e., non-pilot) solenoids for this reason.

One end user experienced solenoid problems due to poor instrument air quality. Their "fix" was to replace instrument air with nitrogen. Unfortunately, the very dry nitrogen created other problems with the solenoid internals, causing 90% of the solenoids to stick. As covered in Chapter 13, one needs to be very careful when making changes. Changes made with the best of intentions (as most are) can have serious side effects.

9.7 Installation Concerns

Field device installation considerations should include:

- environment (temperature, vibration, shock, corrosion, humidity, EMI/RFI)
- on-line testing (if required)
- maintenance requirements
- accessibility
- local indicator
- freeze protection (if necessary)

Refer to Chapter 11 for more details regarding installation issues.

9.8 Wiring of Field Devices

Common wiring failures are:

- open/short
- ground fault
- noise/induced voltages

The following are general guidelines aimed at minimizing wiring problems:

- The standards state that "each field device shall have its own dedicated wiring" to the logic system, unless a suitable bus that meets the performance requirements is available.

- As of the time of this writing (spring 2005), there are no safety-related fieldbuses in widespread use in the process industries, although several are under development (e.g., FOUNDATION Fieldbus) and several are used in other industries (e.g., Profisafe, AS-i). The ISA SP84 committee currently has a task team developing a high-level safety fieldbus specification. The committee fully recognizes that safety fieldbus will be available for the process industries in the near future, hence, the exception statement in the bullet point above.

- It's a relatively common practice to connect several discrete input switches to a single input channel of a logic system in order to save wiring and input module costs. One drawback of such an arrangement however, is that it's more difficult to troubleshoot and diagnose problems. Connecting a single output to several final elements is also relatively common and may be acceptable as long as all the devices are part of the same function.

- Each field input and output should be fused or current-limited. This can be done either as a part of the logic system I/O modules or through the use of external fuses.

- A general recommendation would be to segregate cables, conductors, and junction boxes associated with SISs from all other control and/or instrumentation wiring and clearly label them as part of an SIS.

- Pay attention to the inductance, capacitance, and length of cable runs. Depending upon the wire gauge and the distance involved, induced voltages may occur that could prevent I/O circuits dropping out.

Summary

Field devices include sensors, final control elements, field wiring and other devices connected to the input/output terminals of the logic system. These devices can account for approximately 90% of the potential problems associated with Safety Instrumented Systems. Redundancy and/or diagnostics can be used to improve the performance of field devices.

References

1. ANSI/ISA-84.00.01-2004, Parts 1-3 (IEC 61511-1 to 3 Mod). *Functional Safety: Safety Instrumented Systems for the Process Industry Sector.*

2. *Guidelines for Safe Automation of Chemical Processes*. American Institute of Chemical Engineers - Center for Chemical Process Safety, 1993.

3. Cusimano J. A. "Applying Sensors in Safety Instrumented Systems." *ISA TECH/EXPO Technology Update*. Vol. 1, Part 4 (1997). pp.131-137. Available at www.isa.org (Retrieved 6/28/2005 from source)

4. Faller, R. *Safety-Related Field Instruments for the Process Industry*. TÜV Product Service GMBH, 1996.

10
ENGINEERING A SYSTEM

Chapter Highlights

"Everything should be made as simple as possible, but no simpler."

— *Albert Einstein*

Chapter 7 dealt with the pros and cons associated with various logic system technologies. Designing an SIS, however, involves much more than just choosing a logic box. Chapter 9 dealt with field device issues. This chapter will deal with other management, hardware, and software considerations. Some aspects of system design are impossible to quantify in a

reliability model (the topic of Chapter 8), yet they can have a profound impact on system performance.

The goal of many engineers and companies is to develop a "cookbook" for system design. The senior engineers, with their experience and knowledge, will write down rules and methods to be followed by the junior engineers. One must be cautious, however; one set of rules or procedures simply can't apply to all systems and organizations.

10.1 General Management Considerations

Section 5 of IEC 61511 covers the management of functional safety. The standard naturally focuses on issues related to safety instrumented systems, not general health and safety issues. IEC 61511 states that policies and strategies for achieving safety need to be identified, along with a means of evaluating its achievement. People and departments need to be identified and informed of their responsibilities. Personnel need to be competent to carry out their assigned activities. Competency is a function of knowledge, experience, and training. Competency requirements need to be defined and documented.

10.1.1 Job Timing and Definition

Many system problems are due to two simple factors: a combination of poor definition and timing. Either the scope of the job is too vague, the specification is incomplete (or non-existent at the time of initial purchase or even system design), and the job was due last week. Incomplete specifications often lead to scope creep, increased costs, delays, and dissatisfaction. The more firmly things can be established and documented up front, the better off everyone will be. Chapter 2 summarized the U.K. Health and Safety Executive findings that the majority of system problems (44%) were due to incorrect specifications (functional and integrity).

10.1.2 Personnel

Once a contract is awarded, it's best not to change the people during a project. While reams of information are no doubt part of every contract, Murphy's law predicts that the one thing that isn't will crop up to bite you. That one thing exists only in someone's head, and when that person is transferred that piece of information goes with him, only to rear its ugly head at the worst possible moment.

No matter how much is put in writing, specifications are always open to interpretation. If the people are changed, the project's direction will be lost while new people reformulate opinions. This may have dire consequences on the delivery and price of a contract, since items may need to be renegotiated.

10.1.3 Communication Between Parties

It's important to establish clear lines of communications between the companies involved. The project manager at company A should be the only person talking to the project manager at company B. If other parties start getting involved, certain items may not get documented, managers may not be informed, and important pieces of information will inevitably fall through cracks, only to cause major headaches later on.

10.1.4 Documentation

Will documentation be provided in the user's standard format, the engineering firm's standard format, or the integrator's standard format? What size will the drawings be? Will they be provided on paper only or in digital format? If so, using which program? Will there be 200 loop diagrams, or one "typical" loop along with a table outlining the information on the other 199 loops? Little things like this can easily be neglected until a point in time at which they can become major points of contention.

10.2 General Hardware Considerations

The design and analysis of a safety system requires the consideration of many different factors such as: technology selection, architecture, failure rates and modes, diagnostics, power, interfaces, security, test intervals, etc. Many of these factors were discussed and analyzed in Chapters 7, 8, and 9. Some have gone so far as to call these factors PIPs (primary integrity parameters). Some people just love acronyms!

10.2.1 Energized vs. De-energized Systems

Most SIS applications are normally energized (de-energize to shutdown). Some applications (typically machine control) are just the opposite (normally de-energized and energize to take action). This difference in system design involves some unique considerations. Systems designed for one, may not be suitable for the other. One can't just use some form of logic inverter. [Ref. 8]

For example, in a normally energized system the healthy logic state will be 1, or energized. A safe failure or removal of a system component would result in a shutdown (logic 0) of a non-redundant system. For a normally de-energized system, the healthy logic state will be 0, or de-energized. A safe failure or removal of a system component in such a system should *not* result in a shutdown (logic 1) of a non-redundant system—but it would if the logic were simply inverted. Such systems must be designed differently.

For normally de-energized systems, output solenoid valves should be supplied with latching mechanisms so that when a trip occurs the valves latch in the proper state. Otherwise, if there is a power loss after a shutdown, or batteries are drained, devices could go back to an abnormal running state, which could result in a hazardous condition.

For normally de-energized systems, alarm fuse terminals should be used. These provide alarms in the event that a fuse goes open circuit. Otherwise, an open fuse in a normally de-energized system would not be recognized, thus creating a potentially dangerous failure that could prevent the system from functioning.

For normally de-energized systems, line monitoring (supervision of circuits) of input and output field cabling is recommended (and required for some systems). Line monitoring of field cabling enables detection of open circuits, which again represent potentially dangerous failures that could prevent a normally de-energized system from functioning properly.

Normally de-energized systems also need to be able to monitor (and alarm) for loss of power.

Normally de-energized systems are generally selected when prevention of spurious trips is paramount. This is typically the case with fire and gas and machinery protection systems. Spurious operation of these systems can cause equipment damage and dangerous situations for personnel.

10.2.2 System Diagnostics

Any SIS, no matter what technology or level of redundancy, requires thorough diagnostics. Safety systems are by their very nature dormant or passive, therefore certain failures may not be self-revealing. Dangerous failures could prevent the system from responding to a true process demand. In the case of fail-safe systems (accepting that nothing is 100% fail-safe), such as certain relay or European design solid state systems, periodic diagnostics may be performed manually (e.g., by toggling circuits on and off to see if they respond properly). For most other technologies, additional automatic diagnostic capabilities must be incorporated in the system design.

Some systems are available off the shelf with very thorough diagnostics. For these systems, virtually no extra user interaction is required. For certain other systems, namely general purpose equipment, self-diagnostics may be very limited. These systems typically *do* require extra user interaction.

For example, electronic components may fail on, or energized. In a normally energized system, such a failure may go unnoticed. This is potentially dangerous, for if there were a real process demand the system would be unable to de-energize. General purpose PLCs typically have little diagnostics in the I/O modules (i.e., low diagnostic coverage). Some users have written papers on this subject and described the extra I/O modules and application programming that were required in order to give the system an acceptable level of diagnostics.

10.2.3 Minimize Common Cause

Common cause failures are single stressors or faults that make redundant systems fail. Obvious examples are power (e.g., simplex power to a triplicated controller), software (a bug will make all channels fail), and back-up switches (if the switch does not work, the dual redundant system will fail). Unfortunately, not all failures are so obvious. Studies in nuclear power plants found that approximately 25% of all failures in power generating stations were due to common cause problems. [Ref. 4] Aircraft have crashed because triplicated control lines were severed. A seven-way redundant government computer system failed when a single fiber optic cable was cut. [Ref. 6]

A vicious cycle begins to appear as redundant systems introduce more complexity, which further exacerbates the common cause problem and can lead to even more equipment being installed. When redundant systems are used, common cause influences tend to be external (temperature, vibration, power surges, RFI) along with inadvertent human intervention. Redundancy is effective against random hardware failures, but not against design or other systematic errors. The most effective methods to reduce common cause problems are to a) physically separate devices, and b) use diverse (dissimilar) devices. If redundant controllers are housed in separate panels in separate rooms, a common failure would be unlikely. If two different pressure transmitters, from two different vendors, based on two different measurement technologies are used, it would be more unlikely for them to have a common problem. One might laugh at the obvious common cause potential introduced when connecting multiple sensors to the same process tap, but it *has* happened on more than one occasion. If someone can, someone inevitably will!

10.2.4 Panel Size and Layout

One of the first documents produced during the design stages of a system are general arrangement drawings. These drawings identify the overall size and weight of the cabinets. Multiple stories have been told of how drawings were produced, reviewed and signed by clients, and systems built and shipped, yet the panels were too large to fit through the control room doors.

Consideration should be given to how personnel will access the system. Technicians should have adequate room to access various parts within the cabinet. Otherwise someone squeezing between two pieces of equipment might accidentally trip a circuit breaker. Adequate lighting should be provided inside the cabinet. Parts should be clearly labeled.

It is very common to locate electrical bus bars toward the bottom of most panels simply due to the fact that most panels are designed for bottom entry cables. This can lead to problems if technicians place their metal lunch buckets or tools on the floor of the panel. Some integrators have been known to locate bus bars toward the ceiling of the panel for this very reason.

10.2.5 Environmental Considerations

It is important to specify the environmental requirements in the system specification. Adequate protection should be provided against temperature, humidity, vibration, electrical noise, grounding, contaminants, etc. For example, one of the authors knows of a system that the user wanted to be software-programmable, but there was no more room in the control building. The panel was to be located outdoors, in South Texas, next to the incinerator it was to control in a Class 1, Division 2 area. The system was supplied in a purged, weatherproof enclosure with vortex air coolers to keep it cool in the summer and a heating element to prevent condensation in cold weather. If the integrator was not made aware of the unique location issues, the system would not have operated for very long.

An end user in the Middle East specifies systems that can operate at 140°F without cooling fans. Not many systems can do this. One can only wonder how many failures the user first experienced before they included this requirement in their specifications.

One of the authors visited a facility where most of the control panels had very large red signs stating, "Do not key radios or walkie talkies within 15 feet of this panel." One can only imagine the problems they experienced that led them to install such warnings. At least one safety system has shut down because a technician working on the panel with an open door simply answered his cell phone!

In general, increasing the temperature of electronics by 10°C decreases the life by approximately 50%. Cabinet ventilation (e.g., fans) should keep the temperature inside the panel to within the manufacturer's specification in order to have a reliable system. Similar consideration must be given to vibration, corrosive gases, etc. There are systems available with modules that specifically have large metal heat sinks, rubber gasket seals, and conformally coated electronics to survive such harsh environments. Such a design naturally comes with an extra cost.

10.2.6 Power

Power, like many other topics, is not a black and white issue. There is no clear cut right vs. wrong way of handling power. Simply put, systems need clean, protected, and regulated power. There should be an isolation transformer to protect against spikes, transients, noise, over/under voltage, etc.

Critical safety systems are often supplied with redundant power sources. There are many ways to achieve this. Each system and/or method will have pros and cons that will be unique to each application. Each system must be evaluated for the intended usage. Will switching to the backup unit truly be bumpless? What may work for one installation may not work for another. Backup batteries, like anything else, will not work if they're not maintained. Can anything truly be "uninterruptible"?! *All* portions of the system must be periodically tested. Many of the above problems can be overcome by using redundant DC power sources. Diode ORing them together eliminates AC switches.

Power supplies should be mounted to allow the best heat dissipation while still allowing for easy access and troubleshooting.

10.2.7 Grounding

Proper grounding is important for a successful installation. Grounding is usually more critical for software/electronic systems than older electric systems. Manufacturer's recommendations must be closely followed. Some systems require a fixed ground. Others require a floating ground. A system designed for one may not operate properly the other way. A partial checklist of items to consider would include corrosion, cathodic protection, static electricity, and intrinsic safety barriers.

10.2.8 Selection of Switches and Relays

Relays should be sized not only for the maximum load (to prevent welding of the contacts—the most dangerous failure in an SIS), but the minimum load as well. If this is not accounted for, the trickle current in the circuit might not be enough to keep the relay contacts clean, and circuits may open years later due to dirty contacts. This can be difficult to diagnose, as it's generally not the first thing one looks for.

10.2.9 Bypasses

Bypasses may be necessary for certain maintenance and start-up activities. Leaving systems in bypass, however, is potentially *very* dangerous. If the process can be hazardous while an SIS function is being bypassed, administrative controls and written procedure should be provided to maintain the safety of the process.

Bypasses may be implemented many different ways. Installing jumper wires is potentially dangerous because there's no external indication that something is actually in bypass. Operators may be wondering why the system isn't working properly and may have no clue that some technician somewhere else has placed something in bypass in order to perform maintenance.

Virtually all PLCs can use "forcing" of I/O to accomplish the same thing as a bypass. The ease with which this can be done in some systems is cause for concern. Leaving I/O points forced, without any indication of such a condition, is potentially very dangerous.

Bypass administration is very important to assure that bypasses are promptly and properly removed. This is similar to hot work permits, a problem that contributed to the Piper-Alpha disaster in the North Sea where over 160 people died. Strict documentation and control are required. Some type of bypass form should be used. This form should have operations knowledge and approval, with time limits stated and renewed on shift changes, and may require different levels of approval.

With many systems, when an input is in bypass, there's no indication at the panel of the true state of the input. How do you know whether the signal is healthy or not before you turn the bypass off? What if the input goes into alarm while it's in bypass? How can you tell? The system should be designed so that even when an input is in bypass, the system *still* provides an indication of the true state of the field device.

Returning to normal conditions after a start-up bypass can be accomplished in a variety of ways. For instance, a flame detector will need to be bypassed in order to start up a burner system. Once a flame is detected, the bypass can be disabled since it's no longer desired. A low level shutdown may need to be bypassed in order to start a unit. Once the level reaches a normal reading, the bypass may be disabled. It may be possible in some applications to disable bypasses using timers.

Serious accidents have been documented in the chemical industry while systems were temporarily taken out of service, or portions were operated manually without knowledge of others. [Ref. 6 and 7]

10.2.10 Functional Testing

The requirements for system testing should be incorporated in the design. How will the system be tested? Will portions by bypassed and tested manually? Will some form of automatic testing be incorporated, even for the field devices? The designer might assume devices will be tested yearly. Maintenance technicians may realize such a requirement is totally unreasonable based upon the facilities normal practices. In addition, the device to be tested may be relatively inaccessible, making testing unlikely. The designer may wish to incorporate partial stroke testing of valves. Operations may be concerned about possible nuisance trips and may, therefore, be totally against such an idea. It may prove useful to include maintenance and operations personnel in a design review to catch such potential problems.

10.2.11 Security

How secure will the system be? Who should have access and who should not? How much of the system should different people be able to access? How will access be controlled?

One of the authors knows an engineer who said he was the only individual, with the only portable PC, responsible for making changes to a certain system. Yet every time he hooked up to it, there was a different version of the program running in the controller than he had on his PC. Other people were making unauthorized changes. How was this possible? Some systems do not have any real security features. If this is the case, management procedures must be enforced to prevent unauthorized access. Unfortunately, procedures can and will be violated on occasion.

One of the authors heard a story of a disgruntled engineer, who from his home computer and modem was able to break into the plant DCS and alter its database. Again, something like this should never be possible with an SIS. The age-old problem, however, is that hindsight is always 20/20. It's obviously a bit more difficult to have foresight and predict events that may never have taken place. One might want to keep a variation of Murphy's law in mind: If the system allows it, and someone can do it, at some point they will—no matter how many warnings they may be given not to.

One of the authors toured a facility where the control engineer expressed pride in their security procedures. Upon touring the control room, the author noticed many of the control cabinets open, keys in the doors, and no one else present in the area. The engineer might have been impressed with their security, but the author wasn't.

Keys may be used to control access, but management procedures must be in place and enforced to control their availability and use. Passwords may allow various levels of access into systems (e.g., for bypassing, forcing, changing software), but password access must also be strictly controlled. Do not leave passwords taped to the operator interface!

10.2.12 Operator Interfaces

Many different devices can be used as an operator interface: CRTs, alarms, lights, push-buttons, etc. All of these may be used to communicate information to operators, such as shutdown action taken, bypass logs, system diagnostics, field device and logic status, loss of energy that impacts safety, failure of environmental equipment, etc. The operation of the system, however, must not be dependent upon the interface, as it may not always be functioning or available.

The following questions should be answered in the requirements specification for every data item to be displayed:

1. What events cause this item to be displayed?
2. Can, and should, the displayed item be updated, and if so, by what events (e.g., time, actions taken, etc.)?
3. What events should cause the data displayed to disappear?

Failure to specify the above criteria can result in specification incompleteness and be a potential source of hazards.

It's important to provide certain information to the operators (although not to the point of overload), yet it's just as important to limit what operators are actually able to do at the interface. For example, if a portion of the system is placed in bypass, what indication is there to alert others?

The operator interface is important, but it should not be critical or required for proper operation. It should not be possible to modify the operation of the safety system from the operator interface. Thought should be given as to what should happen if the interface screens go blank (which has been reported on more than one occasion).

Warning signals should not be present too frequently or for too long, as people tend to become insensitive to constant stimuli.

10.3 General Software Considerations

"And they looked upon the software, and saw that it was good. But they just had to add this one other feature... "

— G. F. McCormick

The quality of the application software developed by the end user is naturally an important issue affecting the overall performance of a system. The specification, design, installation, and testing of application software is not always given the priority and emphasis required. This can result in improper operation, delays, and increased costs.

It is common in the process industries to find that the group responsible for development the basic process control system (BPCS) software is also responsible for developing the safety instrumented system (SIS) software. Identical personnel and procedures are frequently used for the implementation of both the BPCS and SIS software. This may not only result in significant potential common cause problems, but a potentially inadequate software development process.

10.3.1 Software Lifecycle

Many techniques, models, and methods have been used to develop application software. These techniques normally follow a particular sequence or set of steps in order to ensure success. The intent is to get it right the first time, rather than being forced to do it over.

The least formal, and potentially most common, method might be termed "Code and Fix." In such cases the requirements are not well defined or specified. The programmer may be forced to write code and then ask, "Is this what you want? " This iterative process may continue until an apparently workable system evolves. The software ends up being developed based on generic guidelines and the *true* requirements end up being defined *after* much of the code is written. This is obviously an ineffective, time consuming, and expensive approach.

Just as there's an overall lifecycle for the development of safety systems, software should also follow an effective lifecycle development process. Figure 10-1 shows the V-model for software development. Note that this is just one of many different software development models. The V-model describes a top-down approach of design and testing. Requirements need to be defined and verification procedures need to be established, in order to check that *each* requirement (block in Figure 10-1) has been fulfilled.

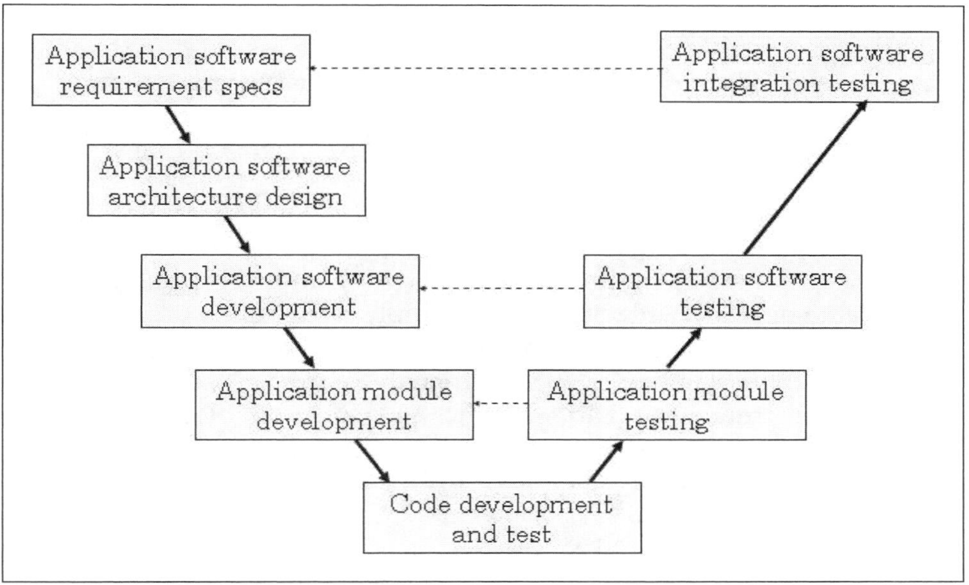

Figure 10-1: Software Development V-model

Examples of items and issues to be considered include:

- **Software requirements**: All possible requirements of the software, including system startup, operations, maintenance, shutdown, alarms, bypasses, etc. need to be clearly documented. The basis for developing the software is the safety requirements specification (SRS). The SRS should contain the logic requirements for each safety instrumented function. If the SRS does not exist or is incomplete, development of effective software will be difficult.

- **Software architecture:** This defines the overall structure of the software that will be used, including language, main program structure, subroutines, interaction between main program and subroutines, standard function blocks, additional or custom functions, etc.

- **Programming:** The coding or actual creation of the software based on the language and architecture defined earlier. There may be single or multiple programs based on the requirements

- **Integration:** Ensuring that all interfaces and interaction with other systems and programs operate as intended.

- **Testing:** Ensuring that the overall requirements have been satisfied.

10.3.2 Program and Language Types

IEC 61511 describes three types of software: application, utility, and embedded. Application software is what the user of the system writes and downloads into the controller. Utility software is used to develop and verify the application program. Embedded software is supplied as part of the programmable system, such as the controller's operating system.

The user is primarily concerned with the development of the application software. The standard describes three development language types: fixed program (FPL), limited variability (LVL), and full variability (FVL). Fixed languages are where the user is simply restricted to the adjustment of few parameters, such as setting the range of a pressure transmitter. Limited variability languages are restricted to predefined, application-specific library functions. Ladder logic and function block programming are examples of limited variability languages. Full variability language programs are more complex and provide a much wider range of functionality, such as C or Pascal.

The use of fixed and limited variability languages is considered suitable for SIL 1-3 applications. The standards do not specify different programming requirements for SIL 1-3 applications. ANSI/ISA-84.00.01/IEC 61511 does *not* cover the use of full variability languages or SIL 4 applications. The user is referred to IEC 61508. It's hoped that full variability languages and SIL 4 applications would not be implemented in the process industries.

Reusing existing software or logic diagrams (e.g., programming a PLC solely based upon the ladder logic diagrams of the relay system it's replacing) may not necessarily increase safety. In fact, it may actually *decrease* it. As stated earlier, most errors can be traced back to the requirements specification. If the requirements were not carefully reviewed during the initial application, faults may still be present. In addition, hazards associated with the new system could not have been considered when the original system was installed (e.g., a PLC is not fail-safe in the same manner as relays).

One of the most common programming languages for software-based safety systems is ladder logic. Ladder logic is relatively simple and generally makes sense to the person who wrote it (at least for that week). Unfortunately, the person who has to make changes to the system years later will probably not be the same person who originally programmed it. Unusual and custom programming tends to be difficult to maintain. This is not imply to it should never be done. It is possible to fully annotate ladder logic programs with detailed comments, but that will not guarantee its clarity.

Different languages are available for programming software-based systems. As stated, ladder logic is probably the most common language. It was the natural outgrowth from hardwired relay-based systems. Ladder logic, however, may not be considered the most appropriate language for some other kinds of functionality. For example, sequential or batch operations are more easily handled by another language: sequential function charts. Math functions may be easier to implement using structured text. The IEC 61131-3 standard covers five languages: ladder logic, function blocks, sequential function charts, structured text, and instruction list. Using a system that implements multiple languages allows various functions to be done in the most appropriate language. This may be easier and less costly than trying to squeeze things into a language that may end up being somewhat awkward.

Standards state that software should be documented in a clear, precise and complete manner. That's very easy to say, but not always very easy to do. Many application programs are developed from paragraphs of descriptive text. Unfortunately, text can be interpreted many different ways. For example, how many pictures can you imagine of three simple words; "man eating lion"? That could be a man at a restaurant eating a steak. A lion steak. Some application programs are developed from logic requirements documented in a cause and effects matrix. However, if the PLC being used must be programmed in ladder logic, someone must convert the logic from the matrix into ladder. Interpretation errors may be introduced as a result.

Some have actually programmed diverse redundant systems by different teams, using different logic boxes and different software packages. Not only does this become a design, debug, maintenance, and documentation nightmare, but history has shown this not to be effective. All the programs must be written to a common specification, and this is where most of the problems lie. Similar problems exist in terms of testing software. The testers may omit the same unusual cases that the developers overlooked.

10.3.3 Quantifying Software Performance

Software does not degrade and fail the way mechanical components do. Software is either coded correctly, or it's not. It's usually some unforeseen combination of events that makes software-based systems fail.

Many have tried to quantify the performance of software. This has developed into a highly contentious issue. There are dozens of techniques and they yield very different answers. The safety system standards committees have, therefore, abandoned the thought of trying to quantify the performance of software. Instead, users are urged to follow a methodical lifecycle detailing requirements, development, and testing. Good specification and development practices should lead to the development of good software. Quantifying what many may realize are poor practices does little to improve anything.

10.3.4 Testing Software

One of the advantages in using a programmable system is the ability to test the logic. This can be done either using the programmable electronic system (PES) processor (often referred to as "emulation"), or on the development PC (often referred to as "simulation"). Emulation involves testing using the same processor that the program will normally run on. Simulation involves testing using what is most likely a different processor than is used in the SIS. The difference may be subtle, but it's worth noting. Just because the program runs in one environment is not a guarantee that it will run in another. Many PES vendors provide some form of application program emulation and/or simulation capabilities with their systems. This allows users to test their application logic both during and after application program development.

Software testing is a means of functionally testing PES logic and operator interfaces without field I/O connected. The testing is accomplished prior to field installation with limited resources away from the plant environment. PES software testing is *not* process or plant simulation (although separate simulation packages are available), nor does it involve testing the BPCS logic or field devices.

Some of the benefits of software testing include:

- Cost: both engineering and startup. It's easier and more cost effective to rectify software or programming problems off-site in a controlled environment than it is in the plant prior to startup. The project schedule can be better maintained because problems in the field will be minimized.

- Early bug fixes. No system is perfect and bugs will exist. The earlier they are identified and corrected, the better.

- Fewer surprises during commissioning and startup. The logic has been tested and interested parties have already "kicked the tires." Software and hardware issues can be segregated since there is confidence that the logic has been fully tested. Less finger pointing will occur.

- Simulation/emulation is an excellent training tool.
- Incorporation of early user feedback.

Simulation/emulation can be done in the designer's office on an ongoing basis as the logic is being developed because no physical I/O is required. Only the application development software is required for simulation (i.e., actual PES hardware is not required).

Summary

Designing an SIS involves many choices encompassing management, hardware and software issues. Although many of these items are difficult, if not impossible, to quantify in a reliability model, they are all of vital importance to the proper and continued long term operation of the system.

References

1. ISA-84.01-1996. *Application of Safety Instrumented Systems for the Process Industries*.
2. IEC 61508-1998. *Functional Safety of Electrical/Electronic/Programmable Electronic Safety-Related Systems*.
3. *Guidelines for Safe Automation of Chemical Processes*. American Institute of Chemical Engineers - Center for Chemical Process Safety, 1993.
4. Smith, David J. *Reliability, Maintainability, and Risk: Practical Methods for Engineers*. 4th edition. Butterworth-Heinemann, 1993. (**Note:** 5th [1997] and 6th [2001] editions of this book are also available.)
5. Leveson, Nancy G. *Safeware - System Safety and Computers*. Addison-Wesley, 1995.
6. Neumann, Peter. G. *Computer Related Risks*. Addison-Wesley, 1995.
7. Belke, James C. *Recurring Causes of Recent Chemical Accidents*. U.S. Environmental Protection Agency - Chemical Emergency Preparedness and Prevention Office, 1997.
8. Gruhn, Paul and A. Rentcome. "Safety Control System Design: Are All the Bases Covered?" *Control*, July 1992.

11

INSTALLING A SYSTEM

Chapter Highlights

11.1 Introduction

Although the title of this chapter is "Installing a system," the intent is to cover key activities ranging from the completion of the design to the successful operation of the system. These activities include the following:

- factory acceptance testing (FAT)
- installation and checks
- validation/site acceptance testing (SAT)
- functional safety assessment/pre-startup safety review (PSSR)
- training
- handover to operations
- startup
- post startup activities

The overall objectives are to ensure that safety systems are installed per the design requirements, that they perform per the safety requirements

specification, and that personnel are adequately trained on how to operate and support the systems. The activities listed above are required in order for this to be accomplished effectively.

The examination of over 30 accidents by the U.K. Health and Safety Executive [Ref. 1] indicated that 6% of the accidents were due to "installation and commissioning" problems. This relatively low figure may be somewhat misleading because the potential for major problems is high if safety systems are not installed as designed and are not commissioned per the safety requirements specification. Some potential problems are:

- Installation errors such as improper/incorrect terminations, or a wrong device being installed for a particular service.
- Inadequate testing in order to detect installation and/or possible design errors. This normally occurs if the testing in not well planned, personnel doing the testing are not adequately trained, or the testing procedures and documentation are poor.
- Use of substandard "free issue" components by the installation contractor. Major components are usually well specified. Minor items, such as terminal strips or connectors, need to be of acceptable quality in order to avoid future problems.

- Unapproved or unwarranted changes made to the system during construction, testing, or startup. It's tempting and usually easy to make "temporary" changes in order to get a system operational. All changes must follow a management of change procedure (see Chapter 13).
- Documentation not updated to reflect as-built condition. This often creates problems when system maintenance is required.
- Inadequate training of operations and support personnel. Management needs to commit to ensuring that all personnel involved with the safety system are adequately trained.
- Temporary bypasses or forces not removed after testing or startup. It's easy to bypass or force inputs and outputs (I/O) of programmable systems. Jumpers on terminal strips may remain unnoticed for long periods. This needs to be managed through access control and procedures.

11.2 Terminology

Terms were introduced in recent standards that may be unfamiliar to some. A few of these terms worth clarifying are:

- Validation
- Verification
- Functional safety assessment
- Pre-startup safety review (PSSR)

Validation: ANSI/ISA-84.00.01-2004, Parts 1-3 (IEC 61511-1 to 3 Mod) describes this activity as "demonstrating that the safety instrumented function(s) and safety instrumented system(s) under consideration after installation meets in all respects the safety requirements specification." This step is usually performed once the system is installed. This is sometimes referred to as a site acceptance test (SAT).

Verification: This term is *not* synonymous with validation. Verification is defined as "the activities of demonstrating for each phase of the relevant safety lifecycle by analysis and/or tests, that, for specific inputs, the outputs meet in all respects the objectives and requirements set for the specific phase." Verification therefore ensures that *each phase* of the safety life cycle has been completed satisfactorily. Validation can therefore be one part of the overall verification activities.

Functional safety assessment: This is defined as "investigation, based on evidence, to judge the functional safety achieved by one or more protection layers." This can be done at various points of the life cycle, but at least one assessment needs to be done after installation, commissioning and validation.

Pre-startup safety review (PSSR): In some countries the final functional safety assessment (i.e., after installation, commissioning and validation) is synonymous with pre-startup safety review (PSSR).

11.3 Factory Acceptance Testing (FAT)

Subclause 13.1.1 of ANSI/ISA-84.00.01-2004, Parts 1-3 (IEC 61511-1 to 3 Mod) defines the objectives of factory acceptance testing as, "to test the logic solver and associated software together to ensure it satisfies the requirements defined in the safety requirement specification. By testing the logic solver and associated software prior to installing in a plant, errors can be readily identified and corrected." The standard also provides recommendations for FAT.

Factory acceptance testing usually pertains to the logic system and operator interface(s), irrespective of the logic technology being used. Whether the logic system consists of 20 relays, or a complex programmable electronic system with hundreds of I/O, the system hardware and software should be tested thoroughly before being shipped to the user.

The text below outlines the benefits, participants, what is to be included in the FAT, and how the testing may be done. Additional details regarding FAT can be obtained from the references listed at the end of this chapter.

Benefits:

- The system hardware and software can be reviewed and tested in a controlled, non-stressful environment away from the plant site.
- Any problems encountered can more easily be resolved and corrected because of the resources available at the vendor's site.
- FAT provides excellent training for support personnel.
- The understanding of the personnel increases, and they have the opportunity to clarify and rectify any misunderstandings.
- The team building between the vendor, design, and support personnel is strengthened.

Participants:

The number of participants required depends upon the complexity and size of the system. The responsibilities of each individual should be well defined and understood. As a minimum, the following personnel should participate:

- Vendor representative(s).
- User design representative. This is typically the individual with the overall design responsibility of the SIS. This individual should have the lead role for the FAT, be responsible for coordinating the FAT, as well as the preparation of the testing procedures.
- End user support personnel.

What is tested

- the complete logic system hardware, including I/O modules, terminations, internal cabling, processors, communication modules and operator interface(s)
- auto switchover and redundancy (if appropriate)
- software (both operations and application)

How the system is tested

Testing needs to be based on well-documented procedures that are approved by both the vendor and the user. Testing consists of, but is not limited to:

- visual inspection
- injecting inputs (e.g., digital, pulse, 4–20 mA, thermocouple) and observing the system response
- driving outputs (digital and/or analog)
- creating various failure scenarios to test backup systems
- testing non-interactive logic (i.e., logic not requiring feedback from field devices to operate)

The complete logic system, including the I/O modules, can be wired to switches, 4–20 mA potentiometers, and pilot lights, each labeled as a field device. The testing crew simulates the operation of field inputs by actuating the discrete switches and analog potentiometers. By noting the status of the output lights, they can observe the status of each output and compare it with the logic diagrams (e.g., cause and effect diagrams) in order to verify system operation. Such testing will also reveal any hardware problems. Various forms of similar testing can also be automated using a PC connected either to the SIS controller in isolation, or including field terminations and I/O modules. The PC can run through testing routines faster than manual methods and also automatically documents the results.

Some users have reported being able to pass the responsibility for the FAT onto the vendor/integrator, with a resulting cost savings.

11.4 Installation

Installation issues include sensors, final control elements, field wiring, junction boxes, cabinets, logic system, operator interface(s), alarms and all other hardware associated with the SIS.

General installation requirements:

- The SIS installation is often part of the overall instrument/electrical installation and may be handled by the same contractor and workforce. One may consider splitting the scope of the SIS work separately from the rest of the instrument and electrical work. This will minimize potential common cause installation problems and reinforce the criticality and special requirements of the SIS (e.g., testing, training, etc.).
- Ensure that the design package to the contractor is complete and accurate. The training and experience of the contractor is important.
- All devices should be installed per the manufacturer's recommendations.
- All equipment and installations must comply with all code and statutory requirements in effect at the local site. The contractor must understand these requirements and ensure compliance.
- All free issue materials supplied by the contractor should be of suitable quality for the intended service. Detailed specifications for these items are not always established.
- All devices should be installed in a manner that allows easy access for maintenance and testing.
- The installation contractor should not make any changes to the calibration or settings of the field devices.
- Care must be taken to protect all field devices from physical and/or environmental damage prior to installation.
- The contractor should not make any changes or deviations from the design drawings without written approval. Management of change procedures must be followed (see Chapter 13). Changes should be recorded on a set of as-built drawings.

11.4.1 Installation Checks

Installation checks ensure that the SIS is installed per the detailed design and is ready for validation. The activities confirm that the equipment and wiring are properly installed and the field devices are operational. The installation checks are best completed by separating the work into two distinct phases.

1. **Field device and wiring checkout.** This is a check on how the field devices are physically installed, the wiring, wiring continuity, terminations, tagging, junction boxes, etc. The installation contractor usually completes this phase with no power to the system.
2. **Device functional checkout.** The functional checkout of the field devices and logic system after the SIS is powered up. The installation contractor or an independent crew may complete this phase.

These checks are intended to confirm that:

- energy sources are operational
- all instruments have been properly calibrated
- field devices are operational
- the logic solver is operational

Various IEC publications [Ref. 3 and 4] contain procedures and forms to carry out the above checks. These procedures and forms assist in ensuring that the devices are properly installed, that all devices are properly checked, and that proper records are maintained.

11.5 Validation/Site Acceptance Tests (SAT)

Validation is commonly referred to as a site acceptance test (SAT). These checks should only be done after the installation checks have been completed. The main objective of validation is to confirm that the system meets the original requirements specified, including the correct functionality of the system logic. One should also ensure that:

- All equipment has been implemented per the vendor's safety manual.
- A test plan is complete with procedures for testing and documenting results.
- All safety lifecycle documents are complete.

The validation shall be satisfactorily completed prior to the introduction of the hazards the SIS is designed to prevent or mitigate. Validation shall include, but not be limited to, the following:

- The system performs under normal and abnormal operating modes (e.g., startup, shutdown, maintenance, etc.)
- The SIS communicates (where required) with the basic process control system or any other system or network.
- The sensors, logic, computations, and final elements perform in accordance with the safety requirement specification.
- The sensors activate at the set points defined in the safety requirement specification.
- Confirmation that functions perform as specified on invalid process variables (e.g., out of range).
- The proper shutdown sequences are activated.
- The SIS provides the proper annunciation and operational displays.
- Computations are accurate.
- Total and partial resets function as planned.
- Bypass and bypass reset functions operate as planned.
- Manual shutdown functions operate as planned.
- Diagnostic alarm functions perform as required.
- Test intervals are documented in maintenance procedures consistent with SIL requirements.
- SIS documentation is consistent with actual installation and operating procedures.

11.5.1 Necessary Documentation

The documentation needed to support the validation depends upon the complexity of the system and what documents were originally prepared by the design team. Because the complete system is being tested, no single drawing will suffice. It's essential that detailed procedures be prepared and followed. The following documentation is usually required to support the validation:

- validation checkout procedures
- a copy of the safety requirement specification
- PES program listing/printout

- a block diagram of the overall system
- a list of inputs and outputs complete with physical addresses
- piping and instrument diagrams
- instrument index
- specification sheets for all major equipment including manufacturer, model, and options
- loop diagrams
- electrical schematics
- BPCS configurations for any SIS inputs or outputs
- logic diagrams (e.g., cause and effect or Boolean diagrams)
- drawings to indicate locations of all major equipment
- junction box and cabinet connection diagrams
- drawings to indicate interconnections and terminations of all wires
- pneumatic system tubing diagrams
- vendor equipment documentation, including (but not limited to) specifications, installation requirements, and operating manuals

11.6 Functional Safety Assessment/Pre-startup Safety Review (PSSR)

Subclause 5.2.6.1.1 of ANSI/ISA-84.00.01-2004, Parts 1-3 (IEC 61511-1 to 3 Mod) states, "A procedure shall be defined and executed for a functional safety assessment in such a way that a judgement can be made as to the functional safety and safety integrity achieved by the safety instrumented system. The procedure shall require that an assessment team is appointed which includes the technical, application and operations expertise needed for the particular installation." As a minimum, at least one assessment needs to be completed prior to hazards being present. The membership of the assessment team must include at least one senior competent person not involved in the project design team. Some of the key requirements outlined in the standard are:

- the hazard and risk assessment has been carried out;
- the recommendations arising from the hazard and risk assessment that apply to the safety instrumented system have been implemented or resolved;

- project design change procedures are in place and have been properly implemented;
- the recommendations arising from the previous functional safety assessment have been resolved;
- the safety instrumented system is designed, constructed, and installed in accordance with the safety requirement specification, and any differences have been identified and resolved;
- the safety, operating, maintenance, and emergency procedures pertaining to the safety instrumented system are in place;
- the safety instrumented system validation planning is appropriate and the validation activities have been completed;
- the employee training has been completed and appropriate information about the safety instrumented system has been provided to the maintenance and operating personnel;
- plans or strategies for implementing further functional safety assessments are in place.

11.7 Training

The criticality of safety systems should be well recognized. Personnel need to be trained in order for systems to operate properly. Training for the safety systems should start with the preparation of the safety requirement specification and continue for the life of the plant. Key individuals that require training include operations and support personnel. Training topics should include:

- understanding the safety requirement specification
- configuration and programming of the PES and other electronic devices
- location of SIS equipment in the control room and field
- SIS documentation and location
- special maintenance procedures to be followed
- online testing procedures
- response to upset situations
- response to emergency shutdown situations
- authorization required before an SIS is bypassed
- understanding how the SIS works (e.g., set points, valves that will close, etc.)

- understanding what to do if a diagnostic alarm goes off
- training for maintenance personnel, including determining who is qualified to work on the system

The opportunities and means for effective training are:

- preparation of the safety requirements specification
- factory acceptance testing
- logic emulation/simulation
- classroom training
- site training manuals and systems
- validation
- startup
- ongoing refresher training

Vendor personnel will be called upon to provide final support. Find out who they are, how to contact them, and whether they're knowledgeable on your system and have the capabilities to provide the required support.

OSHA 29 CFR 1910.119 (Process Safety Management of Highly Hazardous Chemicals) also defines requirements for training with respect to safety systems. The requirements of that regulation should be reviewed.

11.8 Handover to Operations

Satisfactory completion of the validation should be the basis for handing the system over to operations. Each safety function should be signed off to confirm that all tests have been completed successfully. Operations must be satisfied that they have been trained on how to operate the system. If there are any outstanding issues, they must be reflected in the documentation with a clear indication as to:

- what impact the deficiency will have on the operation of the SIS
- when it will be resolved
- who has the responsibility for resolution

If the assessment team believes that any outstanding deficiencies have the potential to create a hazardous event, then the startup should be deferred or re-scoped until the issues are resolved.

11.9 Startup

Safety systems are likely to activate more during the initial startup of a plant compared to any other time. Initial startups are one of the most hazardous phases of operation for some of the following reasons:

- Even if the new plant is completely identical to an existing one, there will always be certain aspects that are unique. Only after extended operation and experience will the actual operation be fully understood.
- Changes are usually requested by operations to bypass or modify trip settings during this period. While this is expected, it's essential that change procedures be implemented and rigorously enforced. The need for and impact of the change needs to be thoroughly analyzed (refer to Chapter 13). Bypasses have been known to be in place for months after startup. One common method of bypassing is to force outputs from the PES. Some form of alarm should be utilized to monitor for any forced functions.
- Operations may not be fully experienced on how to handle upset situations and conditions may escalate to a shutdown.
- Some systematic (design and installation) failures may now be manifesting themselves, resulting in shutdowns.
- System components may be within their infant mortality period. Failure rates are highest during this time.
- There is considerable activity during this period, hence the potential for more human errors.

11.10 Post Startup Activities

The main activities remaining after startup are documentation, final training, and resolution of deficiencies. A schedule should be prepared documenting who will be responsible for each activity along with a timeline. A more comprehensive list of remaining post startup activities includes the following:

- preparation of as-built drawings
- any special maintenance documentation and/or procedures
- list of deficiencies
- training of remaining operations and support personnel
- establishing any necessary online testing programs
- establishing any necessary preventive maintenance programs
- periodic review of system diagnostics

Summary

This chapter reviewed activities from the completion of system design to successful operation. Key activities include factory acceptance testing (FAT), installation, site acceptance testing (SAT), functional safety assessments/pre-startup safety reviews (PSSR), and startup. The overall objective of these activities is to ensure that systems are installed and perform per the safety requirement specification, and to ensure that personnel are adequately trained on how to operate and support the systems.

References

1. *Out of Control: Why control systems go wrong and how to prevent failure.* U.K. Health & Safety Executive, 1995.
2. ANSI/ISA-84.00.01-2004, Parts 1-3 (IEC 61511-1 to 3 Mod). *Functional Safety: Safety Instrumented Systems for the Process Industry Sector.*
3. *Guidelines for Safe Automation of Chemical Processes.* American Institute of Chemical Engineers - Center for Chemical Process Safety, 1993.
4. IEC/PAS 62381 ed.1.0 en:2004. *Activities during the factory acceptance* test (FAT), site acceptance test (SAT), and site integration test (SIT) *for automation systems in the process industry.*
5. IEC/PAS 62382 ed. 1.0 en:2004. *Electrical and instrumentation loop check.*

12

FUNCTIONAL TESTING

Chapter Highlights

12.1 Introduction

Functional testing must be carried out periodically in order to verify the operation of a safety instrumented system (SIS) and to ensure that target safety integrity levels (SIL) are being met. Testing must include the entire system (i.e., sensors, logic solver, final elements, and associated alarms) and be based on clear and well-defined objectives. Responsibilities must be assigned and written procedures must be followed.

Testing should be regarded as a normal preventative maintenance activity. An SIS cannot be expected to function satisfactorily without periodic functional testing. Everything fails, it's just a matter of when. Since safety systems are passive, not all failures are self-revealing. Sensors and valves may stick and electronic components may fail energized. Therefore, safety systems *must* be tested in order to find dangerous failures that would prevent the system from responding to a true demand.

Testing can be either automatic or manual and can include hardware and/ or software. Testing software is a controversial issue. One should realize that mere testing alone cannot detect all software errors (e.g., design errors). As was stated in earlier chapters, most software errors can be traced back to the requirements specification. Therefore, testing software against the requirements specification may not reveal all errors. This chapter focuses on periodic manual testing of hardware.

12.2 The Need for Testing

It's not unheard of to encounter safety systems that have *never* been tested since the initial installation. When these systems were eventually tested, dangerous failures have been found, most of them associated with field devices. Each of these failures was capable of contributing to a potential accident. The main reasons this practice still persists are:

1. Safety instrumented systems are often viewed similarly to basic process control systems (BPCS), hence the assumption that all safety system failures will be revealed. In Chapter 3 "Process Control versus Safety Control," the differences between control and safety were emphasized. One of the key differences is the manner in which safety systems fail. Safety systems, unlike control systems, can fail in *two* separate but distinct modes. They may fail safe and cause a nuisance trip, or they may fail dangerously and not respond to an actual demand (please also refer to Section 8.4). Control systems are active and dynamic, therefore most failures are inherently self-revealing. Control systems do not normally experience dangerous failures. Safety systems, however, are passive or

dormant and *not* all failures are inherently self-revealing. The primary objective of functional testing is to identify dangerous system failures.

2. Well-defined criteria for testing may not be established during the design phase. Even if established, the expected testing and frequencies may not be communicated to the maintenance or operations groups responsible for their implementation.
3. It may be recognized that the need for testing exists, but adequate procedures and facilities may not be provided, therefore, testing may not be done. Testing in this case may be deferred to whenever a major maintenance turnaround occurs. The resulting test frequencies may be inadequate.

The above items would be addressed if the lifecycle outlined in Chapter 2 were followed. The lifecycle addresses the procedures and need for functional testing in the specification, design, operating, and maintenance phases.

Functional testing of safety systems is not a trivial task (especially if it's performed online while the unit is still running). A significant amount of training, coordination, and planning are required in order to implement such tests and prevent shutting down a unit unnecessarily during tests. Unplanned shutdowns can occur, which can create additional hazardous situations and production losses. Therefore, one may encounter resistance to testing from various departments. Statement like, "Not on my watch," or "If it ain't broke don't fix it" may be heard.

One secondary benefit of testing is that it increases the understanding and confidence of personnel involved with the system.

OSHA's process safety management regulation (29 CFR 1910.119) requires employers to "establish maintenance systems for critical process-related equipment including written procedures, employee training, appropriate inspections, and testing of such equipment to ensure ongoing mechanical integrity." The OSHA regulation further stipulates that "inspection and testing procedures shall follow recognized and generally accepted good engineering practices. "

To better understand the need and rationale for functional testing, let's consider a very simple high pressure safety instrumented function (SIF) consisting of a pressure transmitter, relay logic, and single shutdown valve, as shown in Figure 12-1.

Assuming no redundancy or automatic diagnostics, the average probability of failure on demand (PFD_{avg}) can be expressed as follows:

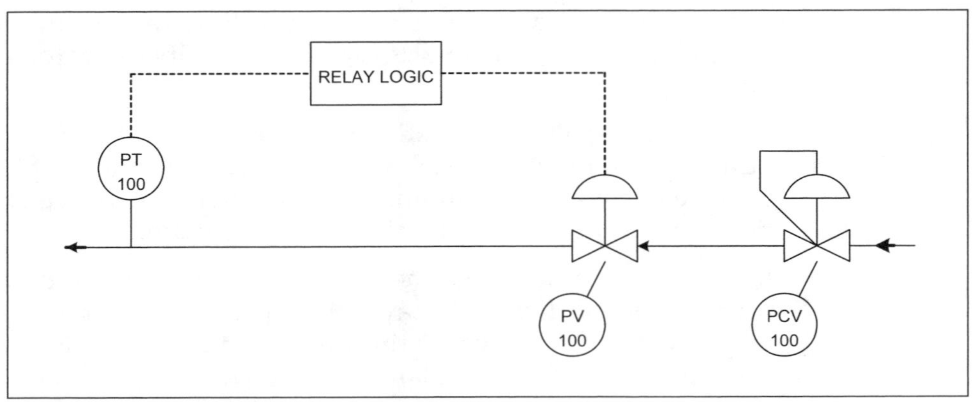

Figure 12-1: High Pressure SIF

$$PFD_{avg} = \lambda_d * TI / 2$$

where:

λ_d = dangerous failure rate of the entire system (i.e., sensor, logic solver, final element)

TI = proof test interval

One can see from the above equation that the proof test interval (TI) is as important a factor in establishing the performance of the system as is the dangerous failure rate of the hardware components. Theoretically, if a system is never tested, the probability of the system failing to respond to a demand approaches 1, or 100%. In other words, if the system's never tested, it's unlikely that it will work when needed.

The above equation takes no credit for automatic diagnostics and also assumes that manual testing is 100% effective. If the testing is not 100% thorough, some dangerous failures may not be identified and may remain undetected for the life of the system. The result is that the PFD_{avg} will be even *higher*. The PFD_{avg} can be calculated for cases where testing effectiveness is less than 100% using the following formula:

$$PFD_{avg} = (E_T * \lambda_d * TI/2) + ((1-E_T) * \lambda_d * SL/2)$$

where:

E_T = Effectiveness of proof test (0 – 100%)

SL = System Lifetime. The system lifetime may be the time the system is completely tested, replaced, or the lifetime of the plant if the system is never fully tested or replaced.

Example:

A shutdown valve has a dangerous failure rate of 0.025 failures per year (or a mean time to failure [MTTF] of 40 years). Assume that a periodic inspection done once a year can detect 95% of failures. The valve is operated for ten years before it's removed from service and replaced. What's the PFD_{avg} for the ten-year operational interval for the following two scenarios?

(a) The valve is tested once every year and replaced every 10 years

(b) The valve is never tested, but only replaced every 10 years

For scenario a):

$$PFD_{avg} = (0.95 * 0.025/yr * 1yr / 2) + (0.05 * 0.025/yr * 10yr / 2)$$

$$= 0.018$$

$$RRF = 55 \text{ (SIL 1 range)}$$

(Risk reduction factor (RRF) = 1/PFD)

For scenario b):

$$PFD_{avg} = 0.025/yr * 10yr / 2$$

$$= 0.125$$

$$RRF = 8 \text{ (SIL 0 range)}$$

The above formula has a low amount of error as long as the mean time to failure is significantly greater than the test or life time interval. Other more detailed calculations should be done when this assumption is not valid.

12.2.1 ANSI/ISA-84.00.01-2004 Requirements for Functional Testing

Subclause 11.8 of the standard states, "The design shall allow for testing of the SIS either end-to-end or in parts. Where the interval between scheduled process downtime is greater than the proof test interval, then online testing facilities are required."

Subclause 16.2.8 states, "Written proof-test procedures shall be developed for every SIF to reveal dangerous failures undetected by diagnostics.

These written test procedures shall describe every step that is to be performed and shall include:

- the correct operation of each sensor and final element;
- correct logic action;
- correct alarms and indications."

Subclause 16.3.1.5 states, "at some periodic interval (determined by the user), the frequency of testing shall be re-evaluated based on various factors including historical test data, plant experience, hardware degradation, and software reliability." Experience may show that the failure rates assumed during the initial calculations were incorrect and test intervals may need to be re-evaluated.

In addition to actual testing, each safety function should be visually inspected periodically to ensure there are no unauthorized modifications and/or observable deterioration (e.g., missing bolts or instrument covers, rusted brackets, open wires, broken conduits, broken heat tracing, and missing insulation). One of the authors observed a case where a very large NEMA 7 (explosion proof) enclosure had only two bolts holding the cover in place. Written on the front of the enclosure in black magic marker were the words, "The engineer who designed this should be shot!" Perhaps the technician did not understand the need for all the bolts to be in place. A simple visual inspection by a qualified person would have revealed the problem. (The author did inform his hosts.)

The user needs to maintain records that verify proof tests and inspections have been completed as required. These records should include at least the following:

(a) description of the tests and inspections performed

(b) dates of the tests and inspections

(c) name of the person(s) who performed the tests and inspections

(d) serial number or other unique identifier of the system tested (e.g., loop number, tag number, equipment number, and SIF number)

(e) results of the tests and inspections (e.g., "as-found" and "as-left" conditions)

In addition to the standard, the ISA SP84 Committee has written an additional technical report on system testing. ISA-TR84.00.03-2002 ("Guidance for Testing of Process Sector Safety Instrumented Functions (SIF) Implemented as or Within Safety Instrumented Systems (SIS)") contains over 220 pages on testing various sensors, logic systems, and final elements.

12.2.2 General Guidelines

When it's not possible to test all components online (e.g., certain valves or motor starters), those devices should be designed with adequate fail-safe characteristics, diagnostics, and/or redundancy so that they can be tested during the turnaround of the plant and still maintain the required performance. Testing facilities should enable the testing to be carried out easily with no interruption to the normal process operation. The design test frequency for the devices should be at least equal to the expected turnaround time of the plant.

Bypass switches for functional testing should be used with caution. Whenever such switches are provided, the input signals for data logging and alarming should *not* be bypassed. Consider the use of an alarm to indicate an active bypass. Safety must be maintained even when a device is placed in bypass. Procedures should be established and followed on the use of bypasses.

Forcing of inputs or outputs within the logic system to simulate inputs or activate outputs should not be considered the equivalent of testing. Forcing should be restricted to maintenance activities when portions of the system may need to be taken out of service. Procedures should be established and enforced in order to assure that bypasses are not abused.

During testing, portions of the safety system may be offline, or its shutdown capabilities may be reduced. This has an impact on overall system performance, as covered in Section 8.9.1. Never testing has a negative impact, and testing too often and purposefully removing a system from service also have a negative impact. One should therefore attempt to optimize the test frequency. In a 1985 study, Lihou and Kabir discuss a statistical technique for optimizing the test frequencies of field devices based on past performance. [Ref. 2]

During the testing and pre-startup phase of any project, modifications may be made to the system in order to get it "operational." These modifications may be setpoint changes, timer settings, installation of jumpers or bypasses, or disabling of shutdown functions. These "modifications" need to be controlled. The system should be retested to verify that it's consistent with the safety requirement specification.

Testing of shutdown valves usually verifies whether they open or close within a specified time. It may be difficult to check for internal mechanical problems. If this is critical, additional facilities may be required (e.g., redundant valves).

12.3 Establishing Test Frequencies

There are no hard and fast rules for determining manual test intervals. One cannot simply say, for example, that SIL 3 systems must be tested monthly, SIL 2 systems quarterly, and SIL 1 systems yearly. Some companies have established standards stipulating that safety systems must be tested at least once per year by trained personnel. Such criteria may be used to establishing certain minimum requirements, but it may lead to oversimplification. Some systems may require more frequent testing, some may require less. The required test frequency is a variable for each system that will depend upon the technology, configuration, and target level of risk.

As covered in Chapter 8, manual test intervals have a considerable impact on overall system performance. It is possible to solve the formulas in order to determine the optimum test intervals in a quantitative manner. Spreadsheets and/or other programs are able to plot system performance based on different manual test intervals. The resulting plots will show the required manual test intervals based upon the desired level of system performance. Many people put in redundant systems specifically so they do *not* have to test them as often.

The frequency of the proof tests can be determined using PFD calculations as shown in the example above and in Chapter 8. Different parts of the system may require different test intervals. For example, field devices, especially valves, may require more frequent testing than the logic solver. Any deficiencies found during testing should be repaired in a safe and timely manner.

When determining the required manual test interval in a quantitative manner, it's important to recognize the impact of the manual test duration (as covered in Section 8.9.1). For example, when a simplex system is tested, it's usually necessary to place the system in bypass (so as to not shut down the process during testing). What's the availability of the system when it's in bypass? Zero. This should be factored into the calculations. A simplex system is reduced to zero, a dual system is reduced to simplex, and a triplicated system is reduced to dual, during the test. Formulas were given in Section 8.9.1 to account for this. This means that there is an *optimum* manual test interval for any system. An example plot is shown in Figure 12-2.

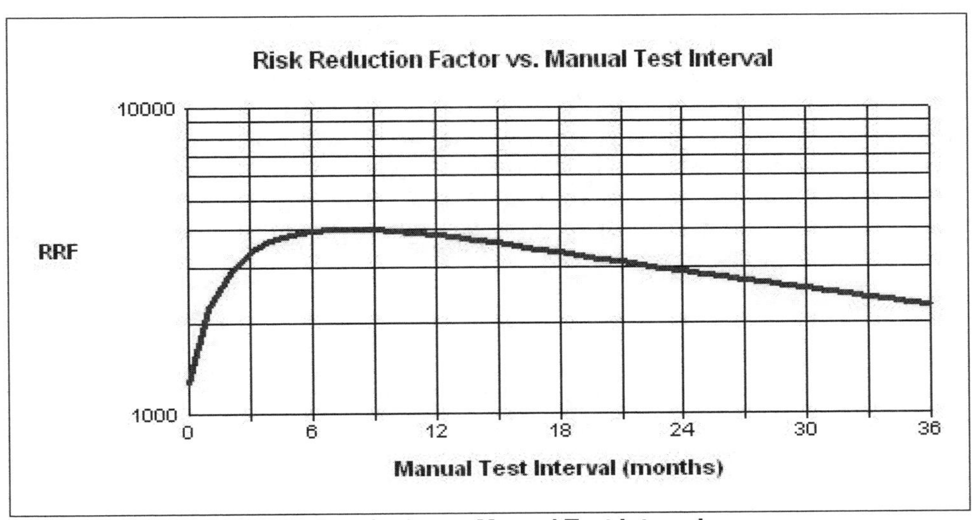

Figure 12-2: Example Showing Optimum Manual Test Interval

12.4 Responsibilities for Testing

It's essential that those involved with functional testing of safety systems be knowledgeable of both the process and the safety system design. Procedures should be written with input from engineering, operations and maintenance. Testing may be performed by the personnel from the process, instrumentation, maintenance, or operations departments. The onus for testing should fall on those ultimately responsible for the safe operation of the plant. Testing facilities and procedures should be established that can be understood and followed by those responsible for the actual testing. In addition to verifying the integrity of the safety system, the testing provides valuable training for those performing the tests, as well as orientation training for new employees witnessing the tests.

12.5 Test Facilities and Procedures

Procedures for testing will vary considerably depending upon the type of device, the testing requirements, and facilities provided for the testing. Testing the system shown in Figure 12-1 might require shutting the unit down. This would be considered offline testing. In order to test the system while maintaining production, online testing capabilities need to be added. For example, in order to provide online testing of the system shown in Figure 12-1:

1. Valves are required to isolate the process connection and pressurize transmitter PT-100.

2. Block and bypass valves need to be installed to test valve PV-100.
3. Figure 12-3 is an example of a similar installation that would allow online testing. In order to test the system, the bypass valve HV-100 would be opened and the signal to transmitter PT-100 changed to simulate a trip. The main trip valve PV-100 should close at the specified set point of PT-100. Some form of feedback (e.g., a limit switch) could be used to verify that the valve PV-100 has in fact closed fully.

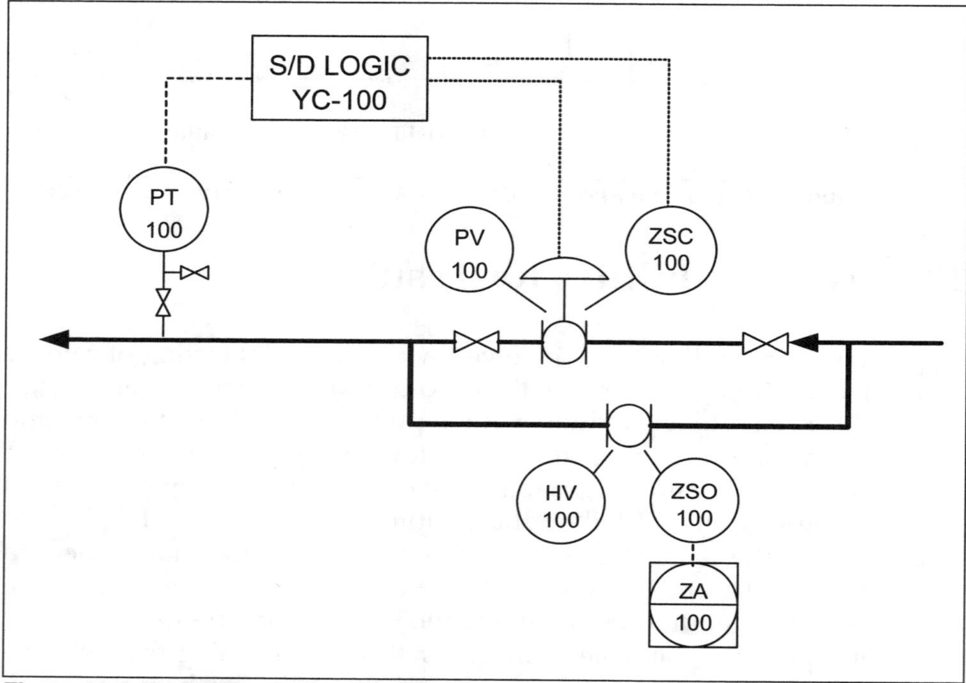

Figure 12-3: Online Testing Capabilities

One way to test a low pressure sensing device would be to close the process isolation valve and vent the signal to the device (if environmentally feasible) until it activates. Similarly, a way to test a high pressure sensing device would be to close the process isolation valve and inject a process signal to the device until it activates. One should have the means to detect that the system responds at the value documented in the requirements specification.

For pneumatic systems, one might consider the use of local indicating pneumatic on/off controllers as an input device. The output signal from the on/off controllers is either 3 or 15 psi. A pressure switch set at 9 psi could act as the input to the logic system. By changing the set point of the on/off controller, the input devices can easily be tested.

If a valve opens for a shutdown, a normally chained opened block valve could be installed upstream of the actuated valve. This block valve could then be closed, and the shutdown valve could be opened for testing. After testing, everything must be returned to their original positions.

The use of manual bypass switches to disable shutdown valves from closing should be discouraged (as they may be misused) but used if it's not practical to install a bypass around the valve.

In some cases it may not be possible to test or shut down a final element (e.g., a compressor, pump, or large valve). In these cases, some form of bypass may need to be installed in order to ensure that the final element won't activate when the sensing element and logic are tested. For example, in the case of a compressor, the final shutdown relay may be energized and the signal from the shutdown relay to the compressor starter bypassed.

If manually-operated bypass switches are to be installed, only one switch should be operated at a time. Active bypasses should be alarmed in some manner. Some bypasses may be necessary in order to permit plant startup (e.g., low level or low flow). These bypasses may need to be controlled by timers that allow the shutdown functions to be re-established after a fixed time. Procedures must be developed and followed on the use of bypasses.

Any defect identified during testing must be reported and repairs scheduled based on the criticality of the system. Redundant systems must also be tested in order to assure that all components are fully functional.

12.6 Documentation

All operating, test, and maintenance procedures must be fully documented and made available to personnel. (However, simply having procedures does not guarantee they'll be *read*.) It's a good idea to not only involve engineering, but operations and maintenance personnel as well in the development of test procedures. No one knows better than they how things actually can and need to be done. The language and terms used should be understood by all parties. Each shutdown function should have a separate test procedure. All testing should leave a paperwork trail, one that can be audited by an independent person or group at a later time.

12.6.1 Sample Test Procedure Documentation

Procedures for testing should contain and list information required by the individuals carrying out the testing. A step-by-step list of actions required to carry out the test should be clearly documented.

An example of key items to be included in a test procedure are as follows:

****

Trip Test Procedure **Procedure #:**_______________

1.0 Purpose:
Clearly define the purpose and objective of the test.

2.0 Responsibility for testing:
One individual must have overall responsibility for the test. Personnel required to assist the responsible individual should also be listed.

3.0 Tools and other items required
Any special tools and other facilities required must be listed. Tools must be calibrated and in good operating condition.

4.0 Trip check frequency/interval
State the test frequency or interval.

5.0 Hazards
Any special hazards which are part of the test, or which can be created as a result of the test should be well documented.

6.0 Reference information

A simplified sketch or description of the system operation could be part of the procedure. Also, relevant loop drawings, spec sheets, schematics, reference or vendor manuals, or material data sheets should be listed.

7.0 Detailed step-by-step test procedure

The detailed step-by-step test procedure must ensure that (as a minimum) the following is verified:

- *Correct functionality of input, logic, output, reset, and alarm devices.*
- *Speed of response of system is acceptable*
- *"As found" vs. expected values should be documented.*

Results / Comments / Deficiencies

Document the results of the test. Any comments or deficiencies identified during the test should be listed.

Trip Test Completed By: ___________________ **Date:** ___________

****

Summary

The testing of safety instrumented systems needs to be carried out in order to verify that they work and to assure their integrity (performance). Testing is also a requirement of various standards, recommended practices, guidelines, and government regulations. Testing must include the complete system (i.e., sensors, logic solver, final elements and associated alarms) and be based on clear and well-defined objectives. Responsibilities must be assigned and written procedures must be followed. Different portions of the system may be tested at different intervals. Optimum test intervals can be determined using simple algebraic calculations. Testing records must be maintained and periodic audits should be performed to check on the effectiveness of procedures and compare information, such as failure rates, with assumptions made during the original design phase of the lifecycle.

References

1. ANSI/ISA-84.00.01-2004, Parts 1-3 (IEC 61511-1 to 3 Mod). *Functional Safety: Safety Instrumented Systems for the Process Industry Sector.*
2. Lihou, D. and Z. Kabir. "Sequential testing of safety systems." *The Chemical Engineer*, December 1985.
3. 29 CFR Part 1910.119. *Process Safety Management of Highly Hazardous Chemicals*. U.S. Federal Register, Feb. 24, 1992.
4. ISA-TR84.00.03-2002. *Guidance for Testing of Process Sector* Safety Instrumented Functions (SIF) Implemented as or Within Safety Instrumented Systems (SIS).

13

MANAGING CHANGES TO A SYSTEM

Chapter Highlights

"The only constant in any organization is change."

13.1 Introduction

Making changes to processes, control systems, safety systems, equipment, and procedures are inevitable. Modifications are required for a variety of reasons, including changes in technology, quality requirements, equipment malfunctions, or safety.

OSHA 29 CFR 1910.119 ("Process Safety Management of Highly Hazardous Chemicals") requires employers to "establish and implement written procedures to manage changes (except 'replacements in kind') to process chemicals, technology, equipment, and procedures; and, changes to facilities that affect a covered process." [Ref. 1]

Subclause 5.2.6.2.2 of ANSI/ISA-84.00.01-2004 states that "Management of modification procedures shall be in place to initiate, document, review, implement and approve changes to the safety instrumented system other than replacement in kind (i.e., like for like)."

This chapter focuses on management of change (MOC) as related to safety systems with specific reference to the OSHA and ANSI/ISA-84.00.01-2004 requirements. The MOC procedures are required to ensure that all changes are made in a safe, consistent, and well-documented manner.

13.2 The Need for Managing Changes

Changes made to a process, no matter how minor, may have severe safety implications. Take one major process accident—the Flixborough explosion in England—as an example.

In June 1974, there was a massive explosion at the Nypro factory in Flixborough that fundamentally changed the manner in which safety is managed by all industries.

The Nypro plant produced an intermediate used to make nylon. The unit that exploded oxidized cyclohexane with air to produce cyclohexanone and cyclohexanol. The process consisted of a series of six reactors, each holding 20 tons, and each partly connected with a flexible joint. Liquid overflowed from one reactor to the next with fresh air added to each reactor. A crack and leak was discovered on one of the reactors and it was removed for repair. A temporary pipe bypass link was installed in its place in order to maintain production.

A qualified engineer was not consulted. The only layout made of the change was a full-scale chalk drawing found on the workshop floor. The system was pressure-tested once installed; but it was a pneumatic rather than a hydraulic test, and it was conducted to 127 psig rather than the relief valve pressure of 156 psig. The design violated British standards and guidelines.

The temporary pipe performed satisfactorily for 2 months until there was a slight rise in pressure. It was less than the relief valve pressure but caused the temporary pipe to twist. The bending movement was strong enough to tear the flexible connection and eject the temporary pipe leaving two 28″ openings.

The resulting gas cloud, estimated to contain 40 to 50 *tons* of material, found a source of ignition in the plant and ignited. The resulting explosion resulted in 28 deaths and 36 injuries, and was estimated to be equivalent to 15 to 45 tons of TNT. The plant was devastated, buildings within 1/3 of a mile were destroyed, windows were shattered up to 8 miles away, and the explosion was heard 30 miles away. Flames from the resulting fire continued to burn for 10 days, hindering rescue work at the site.

The main lesson learned is that *all* modifications must be designed and reviewed by qualified personnel. The men installing the pipe did not have the expert knowledge required. As the world-renowned safety expert Trevor Kletz said, "they did not know what they did not know."

Stringent management of change procedures should be followed to analyze the impact of any and all changes. What may seem to be insignificant changes to one person may, in fact, represent considerable risk. This requires input from people who understand the instrumentation, the control system, the process, and the mechanical/structural systems. It's important to leave a documented, auditable trail.

In order to make any system changes, it's important that the *original* safety design basis be reviewed. Engineers of both hardware and software need to specify the constraints, assumptions and design features, so that maintainers do not accidentally violate the assumptions or eliminate the safety features. Decision makers need to know *why* safety-related decisions were made originally so they don't inadvertently undo them. [Ref. 3] In other words, don't change anything (e.g., a procedure, a piece of equipment, etc.) unless you know *why* it was originally installed or done that way.

Changes can inadvertently eliminate important safety features or diminish the effectiveness of controls. Many accidents have been attributed to the fact that the system did not operate as intended because changes were not fully analyzed to determine their effect on the system. [Ref. 3]

13.3 When Is MOC Required?

MOC would apply to all elements of the process and safety system, including the field devices, logic, final elements, alarms, users interface, and application logic. MOC procedures should be followed for changes made to:

- the process
- operating procedures
- new or amended safety regulations
- the safety requirement specification
- the safety system
- software or firmware (embedded, utility, application)
- correcting systematic failures
- the system design as a result of higher than expected failure rates
- the system design as a result of higher than expected demand rates
- testing or maintenance procedures

A change in firmware constitutes a potentially significant enough modification to warrant following MOC procedures. Manufacturers have the ability to modify and add functionality of devices through software changes. This may not affect the catalog number of the device but would hopefully result in a change in the revision level of the device. The user would work with the manufacturer to evaluate whether the modified device could be used within their system. If the user finds the modified device suitable, they could initially implement it in one channel of a redundant system, allowing them to check its functionality against a plant-approved prior use device. Once they have enough experience with its performance, they could elevate it to full replacement status.

13.4 When Is MOC *Not* Required?

MOC procedures are normally *not* required for:

- Changes deemed as "replacements in kind." "Replacements in kind" does not have the same meaning with regard to safety system components as replacing an electric motor. When replacing an electric motor, one can install another manufacturer's motor if the power, voltage, frequency, speed, and frame size are the same. However, to replace a safety system component—for example, a transmitter—the failure rate and failure modes of the new device are also critical since they directly affect the performance of the system. For safety systems, "replacements in kind" should be "like for like," or a pre-approved replacement which satisfies the original specification.

- Any changes that are within the safety requirement specification.
- Repair of equipment to its original state after failure.
- Calibration, blowdown, or zeroing of field instruments that are part of the safety system.
- Range changes, alarm, or trip changes within design or operation conditions as outlined in the safety requirement specification.
- Any changes covered by another internal corporate procedure which specifically states that the MOC need not be implemented for the specific change.

13.5 ANSI/ISA-84.00.01-2004 Requirements

Clause 17 of the standard covers safety system modifications. The objectives are to assure that modifications to any safety instrumented system are properly planned, reviewed, and approved prior to making the change, and to ensure that the required safety integrity of the system is maintained.

Prior to carrying out any modifications, procedures for authorizing and controlling changes need to be in place. Procedures should include a clear method of identifying and requesting the work to be done and the hazards which may be affected. An analysis needs to be carried out to determine the impact the proposed modification will have on functional safety. When the analysis shows that the proposed modification will impact safety, one should return to the first phase of the safety lifecycle affected by the modification. Modifications should not begin without proper authorization.

Other considerations that should be addressed prior to any changes to the safety system are:

- the technical basis for the proposed change
- the impact of the change on safety and health
- modifications for operating procedures
- necessary time period for the change
- authorization requirements for the proposed change
- availability of controller memory

- effect on response time
- online versus offline change, and the risks involved

The review of the change shall ensure that the required safety integrity has been re-evaluated, and that personnel from appropriate disciplines have been included in the review process.

All changes to a safety system shall initiate a return to the appropriate phase of the safety lifecycle (i.e., the first phase affected by the modification). All subsequent safety lifecycle phases shall then be carried out, including appropriate verification that the change has been carried out correctly and documented. Implementation of all changes (including application software) shall adhere to the previously established design procedures.

When a shutdown system is deactivated, management of change procedures are required to assure that no other systems or processes will be affected. All decommissioning activities therefore require management of change procedures.

Appropriate information needs to be maintained for all safety system changes. The information should at least include:

- a description of the modification or change
- the reason for the change
- identified hazards that may be affected
- an analysis of the impact of the modification activity on the SIS
- all approvals required for the changes
- tests used to verify that the change was properly implemented and the SIS performs as required
- appropriate configuration history
- tests used to verify that the change has not adversely impacted parts of the SIS that were not modified

Modifications need to be performed by qualified personnel who have been properly trained. All affected and appropriate personnel should be notified of any changes and be trained with regard to the change.

Section 12 of the standard covers software, including changes. MOC procedures are also required for changes made to software. Similar to changes with hardware, the objective of MOC related to software is to assure that the software continues to meet the software safety requirement specification. Software modifications shall be carried out in accordance with the following additional requirements:

- Prior to modification, an analysis of the effects of the modification on the safety of the process and on the software design status shall be carried out and used to direct the modification.
- Safety planning for the modification and re-verification shall be available.
- Modifications and re-verifications shall be carried out in accordance with the planning.
- The planning for conditions required during modification and testing shall be considered.
- All documentation affected by the modification shall be updated.
- Details of all SIS modification activities shall be available (for example, a log).

Software changes should be treated as seriously as changes to the plant or process and be subjected to similar control. Any changes to the software should require re-testing of the complete logic associated with the changes.

An offshore user had a PLC-based safety system. There were a number of logic functions that would create a platform-wide shutdown, as shown in Figure 13-1.

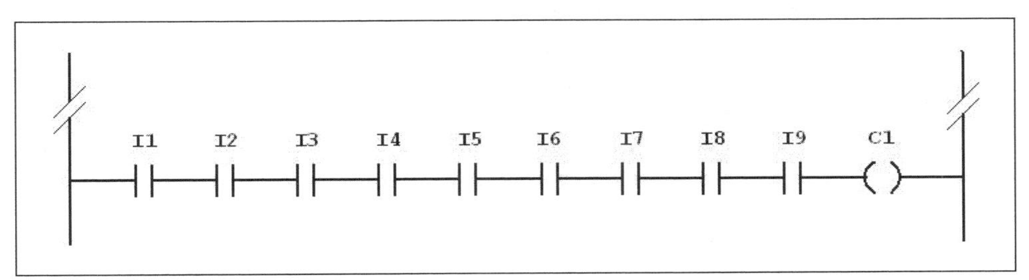

Figure 13-1: Original Ladder Logic

At one point they needed to add a single new logic function that would also cause a platform-wide shutdown. They gave the programming assignment to a relatively inexperienced programmer. The final rung of ladder logic had numerous contacts that would cause a total shutdown (the final coil). In this particular case, there was no room left for any new contacts on that line. "No problem!" thought the engineer, "I'll simply add a new line and repeat the coil on the second line." The resulting logic is shown in Figure 13-2.

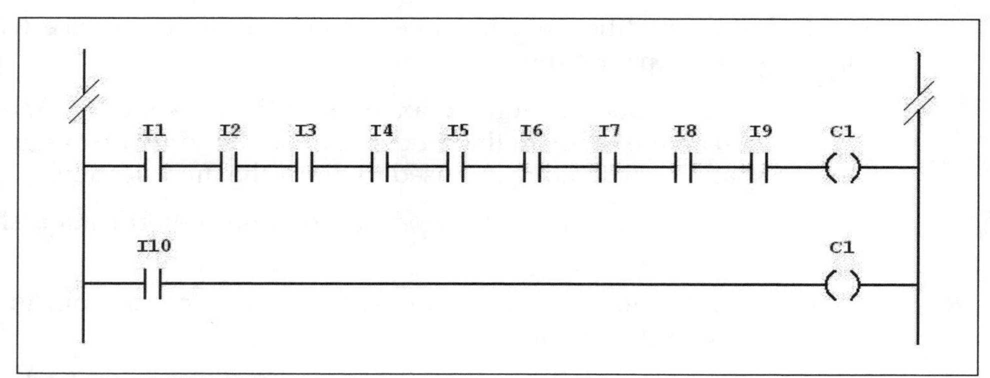

Figure 13-2: Modified Ladder Logic

The user tested the software. Naturally, they only tested the revision, which functioned as desired. They did *not* test any of the original logic. Anyone who understands ladder logic will see the conflict.

PLCs do not immediately write to outputs during the scanning of logic. Variables are held in memory during the logic scan and *afterward* write to the outputs. It's simply a bad programming practice to write to an output in more than one place in a program. One line of logic may de-energize an output, yet another may re-energize it. All this happens in memory *before* actually writing to any outputs. In other words, in ladder logic, the last line wins. (In this case, the original line may call for a shutdown, but the final line will reset the output condition in memory back to "normal." The output will *not* de-energize as intended.) This problem was revealed when a government inspector happened to be visiting the platform. He asked to see a platform shutdown test. The operators were amazed when the system didn't work as intended. *"But we tested it!"* They tried another shutdown function. That didn't work either. The inspector shut the platform down and the owner faced a considerable fine.

13.6 MOC Procedure

The management of change procedure should include the following:

Items to be included in MOC procedures	Comments
How to initiate or request change	Change approval forms need to be developed and included with the MOC procedure. The technical basis for the change should be included in the form. This form is usually generic for all site changes. Both temporary and permanent changes should be included.

Approval of change request	The authorization/approval process for the change should be established.
Responsibilities	Identify the individual(s) responsible for: Approving the change request Identifying the need for implementing a MOC procedure and Who is to ensure that the procedure will be followed
Hazard assessment	The individuals involved in the hazard assessment may not be the same as the original review team. These individuals should be identified. More operational and maintenance personnel are usually involved in the assessment. The techniques required for the assessment must be identified based on the nature, size, and impact of the change.
Documentation	What documentation has to be updated? How will the documentation be updated?
Review and sign off	Who will review the updated documentation and sign off?
Training	Everyone affected by the change needs to be appraised and trained
Construction	When and how will the changes be made in the field?
Check out	A separate crew should be responsible for the checkout based on well-documented procedures.
Startup	Changes should be approved by operations prior to startup.
Final completion	A final report should be signed off and issued upon successful startup.
Auditing and verification	An auditing and verification system should be in place to verify that procedures are being followed.

13.7 MOC Documentation

The following relevant documentation should be updated:

- application logic
- design documentation
- commissioning and pre-startup acceptance testing (PSAT) procedure(s)
- operating procedure(s)
- maintenance procedure(s)
- testing procedure(s)
- safety requirement specifications

Maintaining up to date and accurate documentation for Safety Systems is critical. Inaccurate documentation by itself can create potential hazards. Some key documentation requirements with respect to management of change are:

- All changes to operating procedures, process safety information, and SIS documentation (including software) should be noted prior to startup and updated accordingly.
- The documentation should be appropriately protected against unauthorized modification, destruction, or loss.
- All SIS documents should be revised, amended, reviewed, and approved and be under the control of an appropriate document control procedure.

Summary

Making changes to processes, control systems, safety systems, equipment and procedures is inevitable. Modifications are required for a variety of reasons, including changes in technology, quality requirements, equipment malfunctions, or safety. Regulations and standards state that management of change procedures must be followed for all changes outside of the safety requirement specifications. The procedure must ensure that the changes are:

- *specified* and *designed* by competent individuals
- *reviewed* by an appropriate team capable of addressing the technical basis for the change and identifying the impact of the change
- *inspected* and fully *tested* prior to putting the system back into operation
- fully *documented*
- *communicated* to affected employees so that they are adequately informed and trained.

References

1. 29 CFR Part 1910.119. *Process Safety Management of Highly Hazardous Chemicals.* U.S. Federal Register, Feb. 24, 1992.
2. ANSI/ISA-84.00.01-2004, Parts 1-3 (IEC 61511-1 to 3 Mod). *Functional Safety: Safety Instrumented Systems for the Process Industry Sector.*
3. Leveson, Nancy G. *Safeware - System Safety and Computers.* Addison-Wesley, 1995.

14

JUSTIFICATION FOR A SAFETY SYSTEM

Chapter Highlights

14.1 Introduction

In order to justify the need for a safety instrumented system (SIS), it's important to understand *why* such systems are installed in the first place. Safety systems are used in the process industries to lower the risk to personnel, the environment, loss of production, capital equipment and company image. In fact, it's not unusual to find that the safety integrity level (SIL) determined for some safety instrumented functions (SIF) are higher for economic or environmental reasons than for personnel protection.

The SIL determined for each SIF is usually based on practices developed by a company that follow industry standards (as covered in Chapter 6). The internal practices would normally be calibrated to reflect the amount of risk the company is willing to tolerate. The need to further justify a system may not be required as the justification process is an integral part of their internal practices.

Many companies install safety systems to address only the personnel protection issues and comply with regulatory requirements. In these instances, the SIL determined only reflects the "personnel SIL." With this approach, the assumption is that shutting down the process will address the safety concerns. There usually aren't any other needs for justification—the safety requirements have to be satisfied for legal, moral, financial, and ethical reasons. Justifying a safety system from a purely personnel safety point of view is usually not an issue.

Some companies prefer to have an additional justification step in order to demonstrate that the cost of providing the SIS is justified. If a detailed cost/benefit analysis is completed, the cost of providing the SIS may be shown to be too high and not justifiable compared to not installing the system at all. Justification for providing the system is, therefore, based on carefully analyzing the impact of each of the consequences.

In justifying the need for a safety system, one needs to look at two separate issues:

1. Can the safety instrumented functions be justified based on integrity levels selected to address risk receptors other than personnel (e.g., loss of production, equipment damage, etc.)?
2. The connection between the safety, reliability, and lifecycle costs of the safety system. The safety issue addresses the safety performance of the system based on the risk reduction required. In other words, will the system shut down when required? The reliability issue addresses the nuisance trip performance and the impact nui-

sance trips have on safety and overall costs. In other words, what impact will nuisance trips have on lost production costs and safety (due to the additional risks associated with unplanned shutdown and startup)? Lifecycle cost analysis addresses the total costs of the safety system over its entire life.

Therefore, a safety system can be justified not only for the safety integrity levels selected, but also based on its reliability and overall lifecycle costs.

Another issue facing control system designers is the use of safety instrumented systems, not in addition to, but *instead* of other non-SIS protection layers. Control system designers are sometimes requested to analyze and recommend the possible installation of safety systems to replace other layers because of cost, scheduling, or competitive reasons. A typical example is the installation of high integrity pressure protection systems (HIPPS) to eliminate the need for mechanical relief devices. Some plants have gone through enough upgrades and additions that the original flare systems may no longer be sized to accommodate a total plant-wide upset. It may be less expensive to install pressure protective instrumentation (so relief valves don't lift, thus potentially over-pressuring the flare system) than to change the flare system itself.

This chapter will address the following issues and their impact on safety system justification:

- Failure modes of safety systems
- Justification
- Responsibilities for justification
- How to justify
- Lifecycle cost analysis
- Optimizing safety, reliability, and lifecycle costs

14.2 Safety System Failure Modes

It's been stressed throughout this book that safety systems generally fail in two distinct modes (i.e., safe or dangerous). The impact of these failures on safety system justification can be better understood by further categorizing these two failure modes into four distinct types:

1. *Dangerous undetected failure:* The system has failed dangerously and will not respond to a demand, yet there is no indication of a problem.

2. *Dangerous detected failure:* The system has failed dangerously but the failure *is* annunciated to the operator through the internal self-diagnostics of the system. The system will not respond to a demand until the problem is rectified.
3. *Degraded mode due to a partial failure of a redundant system:* This applies to redundant systems that are fault tolerant. The failure of one channel (safe or dangerous) is annunciated to the operator through the internal self-diagnostics of the system, yet the system *will* respond to a demand.
4. *Nuisance trip failure:* A failure that produces a nuisance or spurious trip (i.e., an unplanned shutdown). The plant has to be restarted once the problem is corrected.

Mode 1: Dangerous Undetected Failure

This is usually considered the worst scenario. The probability of a dangerous undetected failure is the basis for the probability of failure on demand (PFD_{avg}) calculation and the safety integrity level (SIL) of the function or system (as covered in Chapter 8). The SIL determination should be based on the risk tolerance philosophy of the company. This policy should be well established and understood by all parties involved. If this is the case, the SIL determination is unlikely to be challenged by management for personnel protection issues. Whatever safety system architecture or level of redundancy is required in order to comply with the SIL determination is usually readily accepted. The only way to determine whether a dangerous undetected failure has occurred is through manual testing.

Mode 2: Dangerous Detected Failure

This would be the preferred mode if a dangerous failure were to occur. If the mean time to repair (MTTR) of a system was eight hours, and if a demand were to occur on average every six months, then the probability of a hazard occurring within the eight hour repair period would be approximately one in five hundred fifty (1/550).

Upon identifying such a failure, the operator would be expected to follow established procedures. Operating procedures may require the operator to:

1. Monitor the shutdown parameters and the overall process closer and be prepared to take manual action to bring the process to a safe state upon abnormal conditions. In extreme cases, this might require an immediate manual shutdown of the process.
2. Request assistance by support personnel to have the problem rectified. If the problem is not corrected within a predefined time limit,

the system may be designed to shut down automatically. This predefined time limit is usually designated by approval agencies (e.g., TÜV).

3. The level of diagnostic capabilities of the system—including field devices—needs to be established and analyzed as part of the overall system design. There may be additional costs associated with incorporating the level of diagnostics required. This needs to be justified.

Mode 3: Degraded Mode

Redundant systems have the capability to be fault tolerant (i.e., tolerate failures and continue to operate properly) and are widely used whenever both safety and availability/reliability issues need to be addressed. This is obviously the preferred failure mode, since the system will still function while the problem is being rectified. Redundancy increases the initial installation cost, but the overall lifecycle cost of the system may actually be lower. Sections 14.7 and 14.8 show one such example.

Mode 4: Nuisance Trip Failure

Standards such as IEC 61508 and 61511 focus on dangerous failures and performance issues (i.e., will the system shut down when needed) and do not fully address issues associated with safe failures (i.e., nuisance trips). Simply shutting a process unit down whenever there is a system fault is not always the safest or most economical thing to do. The following paragraph in 29 CFR 1910.119 summarizes the concerns about nuisance trips:

"In refining processes, there are occasionally instances when a piece of equipment exceeds what is deemed "acceptable," and interim measures are taken to bring the equipment back into conformance with safe operating parameters. Under (j)(4) it would be mandatory to immediately shut down the entire process upon discovery of such a situation. Shutdowns and startups are inherently dangerous operations, which we try to avoid unless absolutely necessary. In addition, the life expectancy of certain components is directly affected by the number of cycles to which they are subjected. We feel that safety is promoted rather than diminished by keeping shutdowns to a minimum."

It's not unheard of to find portions of safety systems disabled or bypassed when an unacceptable level of nuisance trips have a negative effect on production. Accidents have been reported by OSHA and the EPA as a result of such bypasses. [Ref. 3] To reduce the number of spurious trips, the system may require additional levels of redundancy (i.e., fault-tolerance).

This impacts the cost and complexity of the system and needs to be justified.

14.3 Justification

As mentioned in the introduction of this chapter, there are a number of risk receptors that may benefit from providing safety systems (i.e., personnel, the environment, loss of production, capital equipment and company image). Allocating monies to offset loss of life or injury to personnel can be a very contentious issue. Many companies avoid publishing or using such figures as part of a safety system justification. However, it is possible to quantify the financial impact of not providing a safety system for other reasons. Equipment damage often causes loss of production and these two factors may be combined. Some issues, such as the impact to company image, may be difficult to quantify, but may still be dealt with in more qualitative ways.

The justification for a safety system usually requires the expertise of control system engineering personnel since they have the cost data for safety system installation and operation. Justifications can be based on a cost-benefit analysis. The costs include the engineering, purchasing, installation, operation, and maintenance of a system. The benefits are the cost savings associated with reducing the number of injuries, incidents, and lost production. This chapter will focus on the justification from the following aspects:

1. equipment damage and associated production losses,
2. avoidance of spurious trips.

14.4 Responsibilities for Justification

The responsibilities for justifying a system may be different depending upon whether the system is being considered for protecting equipment and associated production losses, or for the avoidance of spurious trips. (As mentioned previously, justifying a safety system from a purely personnel safety point of view is usually not a problem.) For equipment damage and associated production losses, process and production engineering personnel may need to provide a lead role in the justification.

Issues often arise when recommendations are made to enhance or improve the reliability of a system in order to satisfy the nuisance trip requirements. This often increases the cost and/or complexity of the system. Justifications for such enhancements often face opposition for the following reasons:

- There is a perception by some that this is a production issue, not a safety issue. This is often a false perception.
- The objective of the safety system is to protect the plant. Unwarranted shutdowns due to failures of the safety system may be regarded as a control system issue that should be solved by control system personnel. Production and operations managers do not want to see the plant down for unnecessary reasons and may respond with, "What do you *mean*, your safety system will shut my plant down at least *once every year?!*"
- Project budgets are normally fixed at this stage. There may be reluctance to spend extra dollars unless it can be shown to be well justified.
- The data used for calculating the nuisance trip rate may be challenged.

Justifying a safety system that's above and beyond the SIL requirements is usually the responsibility of the control systems engineer. While the benefits of a more reliable safety system may appear to be obvious, don't assume that everyone will be in agreement. Some common questions will be:

- "The last project didn't require such a complex safety system. Why are you insisting on such complexity and extra cost now?"
- "Can we upgrade the system later if we really have a problem?"
- "What other less costly options have you considered? What's the best alternative?"

There is a need to properly justify the expenditure for any system. All projects are competing for fewer and fewer dollars. The impact of the expenditure on the bottom line will be closely scrutinized.

14.5 How to Justify

The justification for any system is usually based on a financial analysis. Therefore, a review of certain basic financial terms and functions, as well as how the value of money varies with time, is in order.

The future value (FV) of money varies with time and interest rate. If one were to make an annual fixed investment M, the future value of the investment after N years, at an interest rate R, can be expressed as:

$$FV = M\frac{[1+R]^N - 1}{R} \tag{14.1}$$

One can also calculate the present value (PV) of investments made at fixed intervals in the future. If one were to make an annual fixed investment M for N years, at an interest rate R, the present value of the investment can be expressed as:

$$PV = M \frac{1-[1+R]^{-N}}{R} \tag{14.2}$$

When justifying a safety system, we're interested in the present value of money based on losses that can be quantified on an annual basis over a number of years. In other words, one can calculate the impact of a hazardous event and/or nuisance trips on an annual basis, calculate the present value of future losses, and determine the limiting expenditure for the safety system. If the cost is greater than the benefits, the justification is questionable.

Justification Example # 1 - Equipment Damage and Associated Production Losses

A hazardous event may result in an explosion that will cause an estimated $1,000,000 in equipment damage. An analysis showed that the expected frequency of this event is 2.0 x 10^{-3}/year (1/500 per year). A proposed safety system with a risk reduction factor of 100 will reduce the overall risk to 2.0 x 10^{-5}/year (1/50,000 per year). Assuming an expected 20-year life of the safety system and an interest rate of 6%, what level of initial expenditure can be justified for the system?

Solution:

If the safety system reduces the risk from 2.0 x 10^{-3} /year to 2.0 x 10^{-5} / year then, on average, the annual savings will be $2000 - $20 per year, or $1980 ($1,000,000/500 - $1,000,000/50,000). The highest amount of money that can be justified for a safety system will be the present value of an investment that makes 20 annual payments of $1980 at an interest rate of 6%. This is calculated using equation 14.2:

$$PV = \$1,980 * [1 - (1.06)^{-20}]/0.06$$

$$PV = \$22,710$$

The above solution assumes the operation and maintenance costs of the safety system are negligible. In reality, these should also be factored into the calculations.

Justification Example # 2 - Avoidance of Spurious Trips

The total cost of a spurious trip for a unit is estimated at $100,000. The spurious trip rate of the existing safety system is once every 5 years. The safety system can be upgraded at a cost of $70,000 to decrease the spurious trip rate to once every 10 years. If the life of the system is assumed to be 20 years and the interest rate is 5%, is this a justifiable expenditure?

Solution:

The annual cost of spurious trip without an upgrade is $100,000/5 years, or $20,000 per year. The annual cost of spurious trip with the upgrade is $100,000 / 10 years, or $10,000 per year. The potential benefit is $10,000 per year ($20,000 - $10,000).

The annual cost of borrowing $70,000 over a period of 20 years at an interest rate of 5% is obtained by solving equation 14.2 for M.

$$= (\$70,000 * .05) / [1-(1+.05)^{-20}]$$

$$= \$5,617/\text{year}$$

The expenditure is justifiable since the benefit is greater than the cost.

14.6 Lifecycle Costs

One way to justify a safety system expenditure is to complete a lifecycle cost analysis for the various options being considered. The lifecycle costs reflect the total cost of owning the system. By calculating the lifecycle costs, various options can be analyzed in a more quantitative and consistent manner. Table 14-1 describes the predominant costs incurred during the life of a safety system. The list is divided into initial fixed costs (i.e., the costs for designing, purchasing, installing, commissioning, and operating the system), and annual costs (i.e., maintenance and other ongoing costs associated with the system). To some extent, the costs reflect the items listed in the lifecycle model covered in Chapter 2.

Table 14-1: Breakdown of Safety System Costs

Cost Item	Comments
Initial fixed costs	
SIL determination	The costs to complete the safety integrity level (SIL) determination (once the safety review is completed and a safety system is deemed necessary).

Safety requirements and design specifications	The manpower costs to complete the safety requirements specifications, conceptual design, and detailed design specifications.
Detailed design and engineering	Cost for the complete detailed design and engineering.
Sensors	Purchase cost of sensors.
Final elements	Purchase cost of valves and other final elements.
Logic system	Purchase cost of logic system.
Miscellaneous: power, wiring, junction boxes, operators interface.	Costs for other miscellaneous equipment required to install and monitor the safety system.
Initial training	Cost of training for design, operation, and support personnel to design, install, and test system.
FAT/Installation/PSAT	Costs for factory acceptance tests, equipment installation, and pre-startup acceptance tests.
Startup and correction	Most systems require some correction prior to full operation.
Annual costs	
Ongoing training	Ongoing refresher training for operations and support personnel.
Engineering Changes	These costs may be significant due to review requirements and documentation updates.
Service agreement	Programmable logic systems usually required a service agreement with the manufacturer in order to resolve “difficult” problems.
Fixed operation and maintenance costs	Utilities, preventive maintenance programs, etc.
Spares	Critical spares as recommended by vendors.
Online testing	Periodic testing carried out by operations and support personnel.
Repair costs	Costs for repairing or replacing defective modules based on predicted failure rates.
Hazard costs	Cost based on the hazard analysis. The hazard rate is a function of the PFD of the system and the demand rate.
Spurious trips costs	Cost of lost production based on the $MTBF^{spurious}$ of the system.
Present value for annual costs	The present value of the annual costs based on current interest rates and the predicted life of system. These costs are added to the initial fixed costs to obtain the present value of all costs. The PV may be calculated by solving the following equation: $$PV = M\frac{1-[1+R]^{-N}}{R}$$ where M is the annual cost, R is the interest rate and N is the number of years.
Total lifecycle costs	Total costs for the life of the system. This is the sum of the initial fixed costs and the present value for the annual fixed costs.

14.7 Review Example

The following example will be used to develop the lifecycle costs for two possible safety system solutions.

Flammable materials A and B are automatically and continuously fed in a fixed ratio to a reactor vessel by the basic process control system (BPCS). The setpoint of primary flow controller 1 is set by the vessel level controller in order to maintain a fixed level in the vessel. The flow controller for feed A adjusts the setpoint of the flow controller for feed B in order to maintain the fixed ratio.

Figure 14-1 shows the basic process controls.

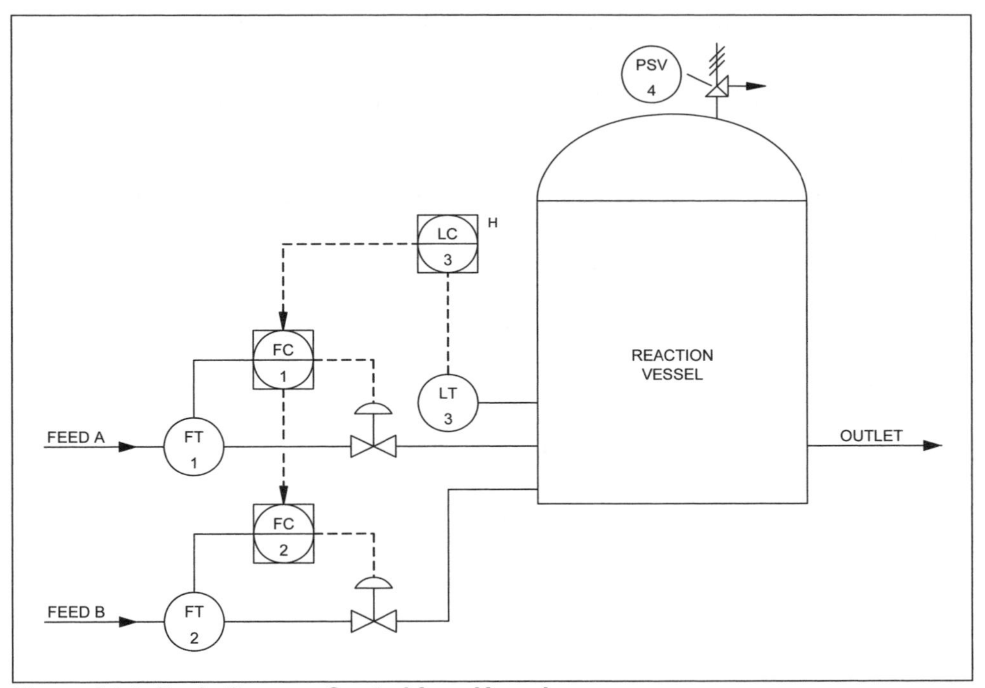

Figure 14-1: Basic Process Control for a Vessel

The following hazardous events were identified in a safety review:

Table 14-2: Summary of Hazard Analysis

Hazard	Cause(s)	Consequence	Likelihood of Occurrence
Release of flammable gas to environment	Failure of BPCS	Fire, explosion $500K loss	Medium
Vessel failure	Failure of BPCS and relief valve	$750K loss	Low
Note: A nuisance trip costs $10,000			

The following safety instrumentation was recommended.

1. Install a high pressure shutdown to close off feeds A and B to the vessel.
2. Install a high level shutdown to close off feeds A and B to the vessel.

Based on the above data, and using the 3-D SIL determination matrix described in Section 6.8 and Figure 6-2, SIL 1 is required for each safety function (pressure and level).

The following safety systems were proposed:

Case 1:

Sensors: Single transmitters

Logic: Relay logic

Valves: Single independent shutdown valves on each line

Case 2:

Sensors: Triplicated transmitters voted 2oo3 (mid value selection)

Logic: Fault-tolerant safety PLC

Valves: Single independent shutdown valves on each line

PFD_{avg} and $MTTF^{spurious}$ Calculations for Case 1

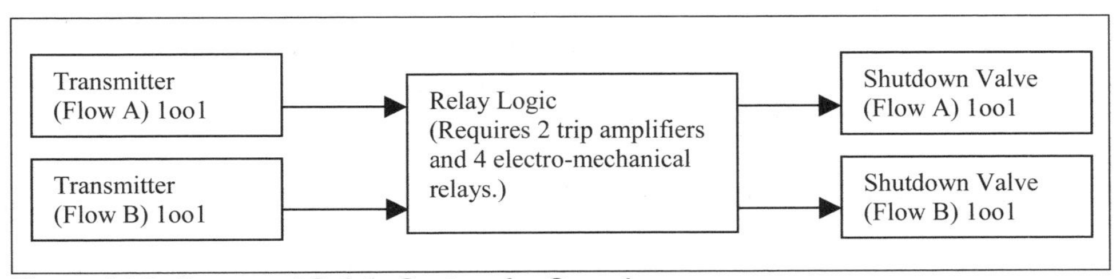

Figure 14-2: Proposed Safety System for Case 1

Failure data: (Failures/year)

Transmitter:	λ_{du} = 0.01	(100 yr MTTF)	λ_s = 0.02	(50 yr MTTF)
Valve and solenoid:	λ_{du} = 0.02	(50 yr MTTF)	λ_s = 0.1	(10 yr MTTF)
Trip amplifier:	λ_{du} = 0.01	(100 yr MTTF)	λ_s = 0.01	(100 yr MTTF)
Mechanical relay:	λ_{du} = 0.002	(500 yr MTTF)	λ_s = 0.02	(50 yr MTTF)

Additional data:

Test interval (TI): 6 months

PFD_{avg} Calculation

The SIL is determined for each individual function. Therefore, the probability of failure on demand (PFD) model should only include one function: 1 transmitter, 1 trip amplifier, 3 mechanical relays and both valves.

PFD_{avg}	=	λ_{du} * TI/2	
PFD_{avg} (Sensor)	=	0.01 * 0.5/2	= 0.0025
PFD_{avg} (Trip amplifier)	=	0.01 * 0.5/2	= 0.0025
PFD_{avg} (Mechanical relay)	=	3 * 0.002 * 0.5/2	= 0.0015
PFD_{avg} (Valve and solenoid)	=	2 * 0.02 * 0.5/2	= 0.0100
PFD_{avg} (Total)			**= 0.0165**

The maximum value allowed for a SIL 1 function is 0.1. Therefore, the above system satisfies the safety requirements. (The risk reduction factor [1/PFD] is 60, which is between 10 and 100 required for SIL 1.)

$MTTF^{spurious}$ Calculation

The mean time to failure, spurious ($MTTF^{spurious}$) calculation should include all components that may cause a shutdown: both transmitters, both trip amplifiers, 4 mechanical relays, and both valves with solenoids. Power supplies are assumed to be redundant and are ignored in the calculation (as non-redundant components will dominate).

$MTTF^{spurious}$ = $1/(\lambda_s)$
$MTTF^{spurious}$ (Sensors) = $1/(2 * 0.02)$ = 25 years
$MTTF^{spurious}$ (Trip amplifier) = $1/(2 * 0.01)$ = 50 years
$MTTF^{spurious}$ (Mechanical relay) = $1/(4 * 0.02)$ = 12.5 years
$MTTF^{spurious}$ (Valve and solenoid)=$1/(2 * 0.1)$ = 5 years
$MTTF^{spurious}$ (Total) 3 years

A nuisance trip may be expected to occur, on average, every 3 years.

While the case 1 system meets the SIL 1 specification requirement, a system with redundant sensors and programmable logic will also be considered.

PFD_{avg} and $MTTF^{spurious}$ Calculations for Case 2

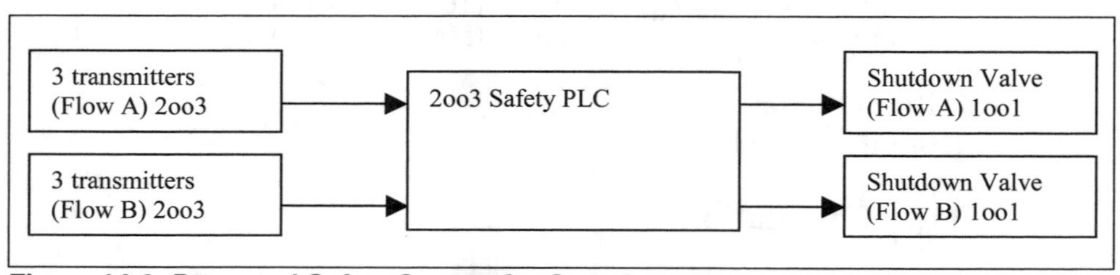

Figure 14-3: Proposed Safety System for Case 2

The failure rate data for the field devices will be the same as in Case 1. The logic system vendor usually provides the safe and dangerous performance of their hardware. Ten percent common cause is included for the redundant transmitters and logic system. Ninety-nine percent diagnostics are assumed for the redundant transmitter arrangement. Refer to Chapter 8 for the various formulas. Testing of the 2oo3 sensors and logic system will be done every 3 years. The valves will still be tested every 6 months.

PFD_{avg} Calculation

The SIL is determined for each individual function. Therefore, the probability of failure on demand (PFD) model should only include one function: 2oo3 transmitters, 2oo3 safety PLC, and both valves. Only the common cause portion of the redundant components needs to be calculated (as it will dominate).

PFD_{avg} (2oo3 sensors) = 0.01 * 0.01 * 0.1 * 3/2 = 0.000015
(failure rate * undetected failure % * common cause % * test interval/2)
PFD_{avg} (2oo3 safety PLC) = (from vendor) = 0.00005
PFD_{avg} (Valve and solenoid) = 2 * 0.02 * 0.5/2 = 0.0100
PFD_{avg} (Total) = 0.01

The maximum value allowed for an SIL 1 function is 0.1. Therefore, this system also satisfies the safety requirements. (The risk reduction factor [1/PFD] is 100, which is the borderline between SIL 1 and SIL 2.)

$MTTF^{spurious}$ Calculation

The mean time to failure, spurious ($MTTF^{spurious}$) calculation should include all components that may cause a shutdown: both 2oo3 transmitter arrangements, 2oo3 safety PLC, and both valves with solenoids. Only the common cause portion of the redundant components needs to be calculated (as it will dominate).

$MTTF^{spurious}$ (2oo3 Sensors) = 1/(2 * 0.02 * 0.1) = 250 years
(1 / quantity * failure rate * common cause %)
$MTTF^{spurious}$ (Safety PLC) = (from vendor) = 200 years
$MTTF^{spurious}$ (Valve and solenoid)=1/(2 * 0.1) = 5 years
$MTTF^{spurious}$ (Total) = 5 years

A nuisance trip may be expected to occur, on average, every 5 years. The valves represent the weak link in Case 2 (both for safety and nuisance trips), as the other portions of the system are fault tolerant. The case 2 system is safer than case 1 and offers fewer nuisance trips. The following lifecycle cost analysis further compares the two systems.

14.8 Detailed Lifecycle Cost Analysis

This analysis is intended to demonstrate the technique for completing the lifecycle cost analysis. *The data is very subjective* and is *not* intended to reflect the actual costs for any particular product or installation technique. Individuals will have to rely on data within their own organization for more accurate numbers.

Table 14-3: Lifecycle Costs for Case 1

	Material $	Labor $	Total cost $	Subtotal $
Initial Fixed Costs				
SIL determination		$2,000	$2,000	
SRS/Design specifications		$5,000	$5,000	
Detailed design and engineering		$30,000	$30,000	
Sensors	$4,000		$4,000	
Final elements	$3,000		$3,000	
Logic system	$10,000		$10,000	
Miscellaneous: power, wiring, junction boxes	$5,000		$5,000	
Initial training		$5,000	$5,000	
FAT/Installation/PSAT	$5,000	$25,000	$30,000	
Startup and correction	$2,000	$8,000	$10,000	
Fixed costs subtotal				$104,000
Annual costs				
Ongoing training		$3,000	$3,000	
Engineering changes	$1,000	$4,000	$5,000	
Service agreement				
Fixed operation and maintenance costs		$1,000	$1,000	
Spares	$2,000		$2,000	
Manual testing		$25,000	$25,000	
Repair costs	$1,000	$500	$1,500	
Hazard costs			$12,000	
Spurious trip costs			$4,000	
Annual costs subtotal				$45,500
PV for annual costs				$567,031
(20 year life, 5% interest rate)				
Total lifecycle costs				**$671,031**

Table 14-4: Lifecycle Costs for Case 2

	Material $	Labor $	Total cost $	Subtotal $
Initial fixed costs				
SIL determination		$2,000	$2,000	
SRS/Design specifications		$5,000	$5,000	
Detailed design and engineering		$30,000	$30,000	
Sensors	$12,000		$12,000	
Final elements	$3,000		$3,000	
Logic system	$30,000		$30,000	
Miscellaneous: power, wiring, junction boxes	$5,000		$5,000	
Initial training		$15,000	$15,000	
FAT/Installation/PSAT	$5,000	$25,000	$30,000	
Startup and correction	$2,000	$8,000	$10,000	
Fixed costs subtotal				$142,000
Annual costs				
Ongoing training		$3,000	$3,000	
Engineering changes	$1,000	$1,000	$2,000	
Service agreement		$2,000	$2,000	
Fixed operation and maintenance costs		$1,000	$1,000	
Spares	$4,000		$4,000	
Manual testing		$15,000	$15,000	
Repair costs	$1,000	$500	$1,500	
Hazard costs			$6,000	
Spurious trip costs			$2,000	
Annual costs subtotal				$36,500
PV for annual costs				$454,871
(20 year life, 5% interest rate)				
Total lifecycle costs				**$596,871**

The Case 2 system has fewer nuisance trips and improved safety compared to Case 1. In addition to the lower manual testing costs (for less frequent testing of the sensors and logic system), the overall lifecycle costs are lower, meaning the Case 2 system is preferred and can be justified.

14.9 Optimizing Safety, Reliability, and Lifecycle Costs

Many chose to focus only on the logic box in the past. Yet simple modeling will usually show that field devices have a much more significant impact, not only on safety performance, but nuisance trip performance and overall lifecycle costs as well. Various options (e.g., redundancy, test intervals, levels of automatic diagnostics, common cause, etc.) can be analyzed and compared based on their long term impact on safety, reliability, and cost.

Reliability modeling and lifecycle cost analysis are very effective tools for optimizing the design of a safety system. Modeling and lifecycle cost analysis allows one to focus on the specific elements of the safety system that require increased attention. By means of the models and the cost analysis, one can better justify system configurations and test intervals based on long term benefits. As shown in the example above, a fault tolerant system may initially cost more but may have lower lifecycle costs due to fewer nuisance trips and reduced testing costs.

This approach can also be used to analyze whether a safety instrumented system can be used as an alternative to other independent protection layers. It's essential that process engineers, as well as other system specialists be involved in the overall analysis.

Summary

Most companies have practices in place to identify the need for safety systems to address personnel safety issues. In such cases, justifying a safety system is usually not an issue.

However, safety systems may be justified not only for personnel safety reasons, but for reliability and total lifecycle cost benefits as well. Other factors that may justify the use of a safety system include protection to capital equipment, loss of production, the environment, and company image. Reliability modeling and lifecycle cost analysis are ideal tools to justify the design and installation of safety systems.

References

1. Goble, W.M. *Control Systems Safety Evaluation and Reliability*. Second edition. ISA, 1998.

2. 29 CFR Part 1910.119. *Process Safety Management of Highly Hazardous Chemicals*. U.S. Federal Register, Feb. 24, 1992.

3. Belke, James C. *Recurring Causes of Recent Chemical Accidents.* U.S. Environmental Protection Agency - Chemical Emergency Preparedness and Prevention Office, 1997.

15

SIS DESIGN CHECKLIST

Introduction

The use of a checklist will not, in and of itself, lead to safer systems, just as performing a HAZard and OPerability study (HAZOP) and not following any of the recommendation will not lead to a safer facility. Following the procedures outlined in the checklist, which are based on industry standards and accumulated knowledge (much of which was learned "the hard way"), should result in safer systems. The checklist is an attempt to list as many procedures and common practices as possible in the hope that, by following a systematic review of the overall design process, nothing will fall through the cracks of an organization and be forgotten.

This checklist is composed of various sections, each corresponding to different portions of the safety lifecycle as described in various standards. Different sections of the checklist are intended for different groups involved with the overall system design, ranging from the user, contractor, vendor, and system integrator. Exactly who has what responsibility may vary from project to project. The checklist, therefore, does not dictate who has what responsibilities, it only summarizes items in the various lifecycle steps.

Why bother? The English Health and Safety Executive reviewed and published the findings of 34 accidents that were the direct result of control and safety system failures. Their findings are summarized in Figure 15-1. Forty-four percent of the accidents were due to *incorrect and incomplete specifications* (functional and integrity), and 20% due to changes made after commissioning. One can easily see that the majority of these issues focus on *user* activities. (Hindsight is always 20/20, foresight is a bit more difficult.) Industry standards, as well as this checklist, attempt to cover *all* of the issues, and not just focus on any one particular area.

This checklist should in no way be considered final or complete. As others review and use it, they are encouraged to add to it (for the benefit of others who may use it in the future). In fact, each section starts with a new page, leaving ample space for your additions and suggestions.

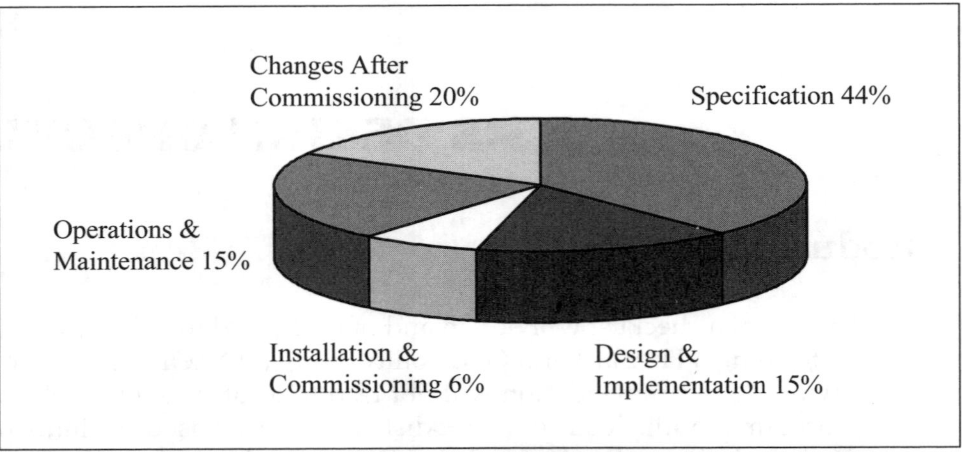

Figure 15-1: Primary Accident Cause by Design Phase, as Published by the U.K. HSE

Overview

This checklist covers the following steps:

- Management Requirements
- Safety Requirements Specification (SRS)
- Conceptual SIS Design
- Detailed SIS Design
- Power & Grounding
- Field Devices
- Operator Interface
- Maintenance/Engineering Interface
- Communications
- Hardware Specification
- Hardware Manufacture
- Application Logic Requirements
- Embedded (Vendor) Software
- Software Coding
- Factory Test
- Installation & Commissioning
- Operations & Maintenance
- Testing
- Management of Change
- Decommissioning

Section 1: Management Requirements

Item #	Item	Circle a choice			Comments
1.1	Have persons or departments responsible for carrying out the phases of the lifecycle been identified?	Y	N	N/A	
1.2	Have persons or departments responsible for carrying out the phases of the lifecycle been informed of their responsibilities?	Y	N	N/A	
1.3	Are persons competent to perform the tasks assigned to them?	Y	N	N/A	
1.4	Is personnel competency documented in terms of knowledge, experience, and training?	Y	N	N/A	
1.5	Has a hazard/risk assessment been performed?	Y	N	N/A	
1.6	Is a safety plan in place that defines the required activities?	Y	N	N/A	
1.7	Are procedures in place to ensure prompt and satisfactory resolution of recommendations?	Y	N	N/A	
1.8	Are procedures in place to audit compliance with the requirements?	Y	N	N/A	

Section 2: Safety Requirements Specification

Item #	Item	Circle a choice			Comments
2.1	Do the safety requirements originate from a systematic hazard assessment? If not, what are the requirements based on?	Y	N	N/A	
2.2	Is there a clear and concise description of each safety related function to be implemented in the SIS?	Y	N	N/A	
2.3	Has the safe state of the process been defined for each operating state of the plant? (e.g., startup, normal operation, maintenance, etc.)	Y	N	N/A	
2.4	Are safety functions defined for each operating state of the plant?	Y	N	N/A	
2.5	Are performance requirements (e.g., speed, accuracy, etc.) defined for each safety function?	Y	N	N/A	
2.6	Has the required safety integrity level (SIL) been determined for each safety function?	Y	N	N/A	
2.7	Are sensor inputs defined with regard to range, accuracy, noise limits, bandwidth, etc.?	Y	N	N/A	
2.8	Are outputs defined with regard to range, accuracy, update frequency, etc.?	Y	N	N/A	
2.9	In the event of system failure, are sufficient information and means available for the operator to assume safe control?	Y	N	N/A	
2.10	Is the operator interface defined in terms of data display, alarms, etc.?	Y	N	N/A	
2.11	Have local or application specific regulatory requirements been considered?	Y	N	N/A	
2.12	Has the operation and implementation of resets been defined for each input and output?	Y	N	N/A	
2.13	Have the operation of bypasses/ overrides been defined for each input and output?	Y	N	N/A	
2.14	Have process common cause considerations (e.g., corrosion, plugging, coating, etc.) been considered?	Y	N	N/A	

Section 3: Conceptual SIS Design

Item #	Item	Circle a choice			Comments
3.1	Are safety functions being handled by a completely separate system from the process control? If not, what is the justification?	Y	N	N/A	
3.2	If multiple functions are being performed within the same logic solver, do the shared components meet the highest SIL requirement?	Y	N	N/A	
3.3	Has the technology and level of redundancy been selected for each safety function? If so, what is it?	Y	N	N/A	
3.4	Have manual test intervals been determined and justified for each safety function?	Y	N	N/A	
3.5	Has the performance of each safety function been analyzed and documented in a quantitative manner in order to see if it meets the safety integrity level (SIL)? If not, what is the justification for the system configuration?	Y	N	N/A	
3.6	Are proven-in-use criteria established for non-certified equipment?	Y	N	N/A	

Section 4: Detailed SIS Design

Item #	Item	Circle a choice			Comments
4.1	Are design documents under control of a formal revision and release program?	Y	N	N/A	
4.2	Has the SIL of the final system been analyzed and documented in a quantitative manner? If not, what is the justification for the system configuration?	Y	N	N/A	
4.3	Are suitable interfaces between field devices and the logic solver defined?	Y	N	N/A	
4.4	Are suitable communication interfaces defined in terms of protocols and information to be exchanged?	Y	N	N/A	
4.5	Are there provisions for future expansion?	Y	N	N/A	
4.6	Are there provisions for incorporating changes as the design proceeds?	Y	N	N/A	
4.7	Is the system fail-safe in terms of:				
	a) Loss of power?	Y	N	N/A	
	b) Loss of instrument air?	Y	N	N/A	
	c) Field cable faults?	Y	N	N/A	
4.8	Can the action of a non-safety function interrupt or compromise any safety functions?	Y	N	N/A	
4.9	Are the safe states of each system component defined?	Y	N	N/A	
4.10	Has the impact of failure of each component in the system been considered, and the required action to be taken defined?	Y	N	N/A	
4.11	Is field I/O power separate from other circuits?	Y	N	N/A	
4.12	Are I/O bypasses incorporated?	Y	N	N/A	
4.13	When an input bypass is enabled, can the state of the sensor still be determined?	Y	N	N/A	
4.14	Are there means for alarming a bypass after a pre-determined time interval?	Y	N	N/A	
4.15	Does the system incorporate manual resetting to restart production? If not, what is the justification?	Y	N	N/A	

Section 5: Power & Grounding

Item #	Item	Circle a choice			Comments
5.1	Are power supplies direct current (DC)? If not, what is the justification?	Y	N	N/A	
5.2	Is a redundant main power source available? If not, what is the justification?	Y	N	N/A	
5.3	Has the impact of power failures been considered?	Y	N	N/A	
5.4	Have the following power concerns been addressed?				
	a) Voltage and current range, including in-rush current?	Y	N	N/A	
	b) Frequency range?	Y	N	N/A	
	c) Harmonics?	Y	N	N/A	
	d) Non-linear loads?	Y	N	N/A	
	e) AC transfer time?	Y	N	N/A	
	f) Overload and short circuit protection?	Y	N	N/A	
	g) Lightning protection?	Y	N	N/A	
	h) Protection against transient spikes, surges, brownouts, and noise?	Y	N	N/A	
	i) Under and over voltage?	Y	N	N/A	
5.5	Have the following grounding concerns been addressed?				
	a) Corrosion protection?	Y	N	N/A	
	b) Cathodic protection?	Y	N	N/A	
	c) Static electricity protection?	Y	N	N/A	
	d) Shield ground?	Y	N	N/A	
	e) Test ground?	Y	N	N/A	
	f) Intrinsic safety barrier grounds?	Y	N	N/A	
	g) Appropriate isolated communications techniques (e.g., communication transformers, fiber optics) between ground planes?	Y	N	N/A	

Section 6: Field Devices

Item #	Item	Circle a choice			Comments
6.1	Is there valid failure rate, failure mode, and diagnostic coverage information for all devices?	Y	N	N/A	
6.2	Have vendors provided recommended manual test intervals and procedures?	Y	N	N/A	
6.3	Will means be available to periodically check the devices for hidden dangerous failures?	Y	N	N/A	
6.4	Are circuits normally energized? If not, is line monitoring (supervised circuits) being incorporated?	Y	N	N/A	
6.5	Does each device have its own dedicated wiring? If not, what is the justification?	Y	N	N/A	
6.6	If smart sensors are being used, are they write-protected?	Y	N	N/A	
6.7	Have minimum, as well as maximum, electrical loads been considered for field I/O circuits?	Y	N	N/A	
6.8	Is feedback available to tell if the final element has moved to its commanded state?	Y	N	N/A	
6.9	Have materials (seals, etc.) been properly selected for the particular application?	Y	N	N/A	
6.10	Does the user have good field experience with the devices in other applications?	Y	N	N/A	
6.11	Are solenoid valves protected from plugging, dirt, insects, freezing, etc.? What measures have been applied?	Y	N	N/A	
6.12	Have the following areas been considered for final elements:				
	a) Opening and closing speeds?	Y	N	N/A	
	b) Shutoff differential pressure?	Y	N	N/A	
	c) Leakage?	Y	N	N/A	
	d) Fire resistance of body, actuator, and impulse lines?	Y	N	N/A	
6.13	Are safety critical field devices identified in some unique manner (e.g., color coding, labeling)?	Y	N	N/A	

Section 7: Operator Interface

Item #	Item	Circle a choice			Comments
7.1	Has failure (loss) of the interface been considered?	Y	N	N/A	
7.2	Are alternate means available to bring the process to a safe state?	Y	N	N/A	
7.3	Are the following shown on the interface:				
	a) Where the process is in a sequence?	Y	N	N/A	
	b) Indication that a SIS action has occurred?	Y	N	N/A	
	c) Indication that a SIS function is bypassed?	Y	N	N/A	
	d) Indication that a SIS component or subsystem has failed or is in a degraded state?	Y	N	N/A	
	e) Status of field devices?	Y	N	N/A	
7.4	Is the update time appropriate for the application under emergency conditions?	Y	N	N/A	
7.5	Have operators been checked for color blindness?	Y	N	N/A	
7.6	Is it possible to change SIS program logic from the operator interface?	Y	N	N/A	
7.7	Do parameters that can be changed have security access protection?	Y	N	N/A	

Section 8: Maintenance/Engineering Interface

Item #	Item	Circle a choice			Comments
8.1	Can failure of this interface adversely affect the SIS?	Y	N	N/A	
8.2	Is there adequate access security? What methods are utilized?	Y	N	N/A	
8.3	Is the maintenance/engineering interface used as the operator interface?	Y	N	N/A	
8.4	Is the maintenance/engineering interface disconnected during normal system operation?	Y	N	N/A	

Section 9: Communications

Item #	Item	Circle a choice			Comments
9.1	Can communication failures have an adverse affect on the SIS?	Y	N	N/A	
9.2	Are communication signals isolated from other energy sources?	Y	N	N/A	
9.3	Has write protection been implemented so that external systems cannot corrupt SIS memory? If not, why?	Y	N	N/A	
9.4	Are interfaces robust enough to withstand EMI/RFI and power disturbances?	Y	N	N/A	

Section 10: Hardware Specification

Item #	Item	Circle a choice			Comments
10.1	Has the physical operating environment been defined and have suitable specifications been set for:				
	a) Temperature range?	Y	N	N/A	
	b) Humidity?	Y	N	N/A	
	c) Vibration and shock?	Y	N	N/A	
	d) Ingress of dust and/or water?	Y	N	N/A	
	e) Contaminating gases?	Y	N	N/A	
	f) Hazardous atmospheres?	Y	N	N/A	
	g) Power supply voltage tolerance?	Y	N	N/A	
	h) Power supply interruptions?	Y	N	N/A	
	i) Electrical interference?	Y	N	N/A	
	j) Ionizing radiation?	Y	N	N/A	
10.2	Are failure modes known for all components?	Y	N	N/A	
10.3	Has the vendor supplied quantitative safe and dangerous system failure rates, including assumptions and component data used?	Y	N	N/A	
10.4	Has the vendor provided diagnostic coverage values for their system?	Y	N	N/A	
10.5	Are logic system components (I/O modules, CPU, communication modules, etc.) all from the same vendor?	Y	N	N/A	
10.6	Has the resulting action of restoring power to the system been considered?	Y	N	N/A	
10.7	Are all I/O modules protected from voltage spikes?	Y	N	N/A	
10.8	If redundant devices or systems are being considered, have measures been taken to minimize potential common cause problems. If so, what are they?	Y	N	N/A	

Section 11: Hardware Manufacture

Item #	Item	Circle a choice			Comments
11.1	Can the vendor provide evidence of an independent safety assessment of the hardware?	Y	N	N/A	
11.2	Does the vendor maintain a formal revision and release control program?	Y	N	N/A	
11.3	Are there visible indications of version numbers on the hardware?	Y	N	N/A	
11.4	Does the vendor have specifications and procedures for the quality of materials, workmanship, and inspections?	Y	N	N/A	
11.5	Are adequate precautions taken to prevent damage due to static discharge?	Y	N	N/A	

Section 12: Application Logic Requirements

Item #	Item	Circle a choice			Comments
12.1	Do all parties have a formal revision and release control program for application logic?	Y	N	N/A	
12.2	Is the logic written in a clear and unambiguous manner that is understandable to all parties?	Y	N	N/A	
12.3	Does the program include comments?	Y	N	N/A	
12.4	Within the logic specification, is there a clear and concise statement of:				
	a) Each safety-related function?	Y	N	N/A	
	b) Information to be given to the operator?	Y	N	N/A	
	c) The required action of each operator command, including illegal or unexpected commands?	Y	N	N/A	
	d) The communication requirements between the SIS and other equipment?	Y	N	N/A	
	e) The initial states for all internal variables and external interfaces?	Y	N	N/A	
	f) The required action for out-of-range variables?	Y	N	N/A	

Section 13: Embedded (Vendor) Software

Item #	Item	Circle a choice			Comments
13.1	Can the vendor provide evidence of an independent safety assessment of all embedded software?	Y	N	N/A	
13.2	Has the software been used in similar applications for a significant period of time?	Y	N	N/A	
13.3	Is the vendor software documented sufficiently for the user to understand its operation and how to implement the desired functionality?	Y	N	N/A	
13.4	Are the results of abnormal math operations fully documented?	Y	N	N/A	
13.5	Are there procedures for the control of software versions in use and the update of all similar systems?	Y	N	N/A	
13.6	For spares which contain firmware, is there a procedure to ensure all modules are compatible?	Y	N	N/A	
13.7	Can software versions in use easily be checked?	Y	N	N/A	
13.8	If errors are found in embedded software, are they reported to and corrected by the vendor, and incorporated into the SIS only after checking and testing the corrected code?	Y	N	N/A	
13.9	Does the manufacturer provide competent technical support?	Y	N	N/A	

Section 14: Software Coding

Item #	Item	Circle a choice			Comments
14.1	Are there standards and procedures for software coding?	Y	N	N/A	
14.2	Are there procedures for documenting and correcting any deficiencies in the specification or design revealed during the coding phase?	Y	N	N/A	
14.3	Are departures from or enhancements to the requirements of the design documented?	Y	N	N/A	
14.4	Is a formal language or some other means taken to assure the program is both precise and unambiguous?	Y	N	N/A	
14.5	Is there a procedure for generating and maintaining adequate documentation?	Y	N	N/A	
14.6	Does the programming language encourage the use of small and manageable modules?	Y	N	N/A	
14.7	Does the code include adequate comments?	Y	N	N/A	
14.8	Are design reviews carried out during program development involving users, designers, and programmers?	Y	N	N/A	
14.9	Does the software contain adequate error detection facilities associated with error containment, recovery, or safe shutdown?	Y	N	N/A	
14.10	Are all functions testable?	Y	N	N/A	
14.11	Is the final code checked against the requirements by persons other than those producing the code?	Y	N	N/A	
14.12	Is a well-established compiler/assembler used?	Y	N	N/A	
14.13	Is the compiler/assembler certified to recognized standards?	Y	N	N/A	

Section 15: Factory Test

Item #	Item	Circle a choice			Comments
15.1	Are there procedures for testing the finished system?	Y	N	N/A	
15.2	Are records maintained of test results?	Y	N	N/A	
15.3	Are there procedures for documenting and correcting any deficiencies in the specification, design or programming revealed during testing?	Y	N	N/A	
15.4	Is testing carried out by persons other than those producing the code?	Y	N	N/A	
15.5	Is software tested in the target system rather than simulated?	Y	N	N/A	
15.6	Is each control flow or logic path tested?	Y	N	N/A	
15.7	Have arithmetic functions been tested with minimum and maximum values to ensure that no overflow conditions are reached?	Y	N	N/A	
15.8	Are there tests to simulate exceptions as well as normal conditions?	Y	N	N/A	
15.9	Have all of the following items been tested?	Y	N	N/A	
	a) Dependence on other systems/ interfaces	Y	N	N/A	
	b) Logic solver configuration	Y	N	N/A	
	c) Operation of bypasses	Y	N	N/A	
	d) Operation of resets	Y	N	N/A	
	e) All functional logic	Y	N	N/A	

Section 16: Installation & Commissioning

Item #	Item	Circle a choice			Comments
16.1	Have personnel received appropriate training?	Y	N	N/A	
16.2	Is there sufficient independence between those carrying out the work and those inspecting it?	Y	N	N/A	
16.3	Have adequate precautions been taken for storage of items during installation?	Y	N	N/A	
16.4	Are installation procedures for all devices sufficient in detail so as not to leave important interpretations or decisions to installation personnel?	Y	N	N/A	
16.5	Has the SIS been inspected in order to reveal any damage caused during installation?	Y	N	N/A	
16.6	Are items such as cabinets, junction boxes, and cables protected from:				
	a) Steam leaks?	Y	N	N/A	
	b) Water leaks?	Y	N	N/A	
	c) Oil leaks?	Y	N	N/A	
	d) Heat sources?	Y	N	N/A	
	e) Mechanical damage?	Y	N	N/A	
	f) Corrosion (e.g., process fluids flowing from damaged sensors to junction boxes, the logic cabinet, or the control room)?	Y	N	N/A	
	g) Combustible atmospheres?	Y	N	N/A	
16.7	Are safety related systems clearly identified to prevent inadvertent tampering?	Y	N	N/A	
16.8	Has the proper operation of the following items been confirmed?				
	a) Proper installation of equipment and wiring?	Y	N	N/A	
	b) Energy sources are operational?	Y	N	N/A	
	c) All field devices have been calibrated?	Y	N	N/A	
	d) All field devices are operational?	Y	N	N/A	
	e) Logic solver is operational?	Y	N	N/A	
	f) Communication with other systems?	Y	N	N/A	
	g) Operation and indication of bypasses?	Y	N	N/A	
	h) Operation of resets?	Y	N	N/A	
	i) Operation of manual shutdowns?	Y	N	N/A	

16.9	Is the documentation consistent with the actual installation?	Y	N	N/A	
16.10	Is there documentation showing the following:				
	a) Identification of the system being commissioned?	Y	N	N/A	
	b) Confirmation that commissioning has been successfully completed?	Y	N	N/A	
	c) The date the system was commissioned?	Y	N	N/A	
	d) The procedures used to commission the system?	Y	N	N/A	
	e) Authorized signatures indicating the system was successfully commissioned?	Y	N	N/A	

Section 17: Operations & Maintenance

Item #	Item	Circle a choice			Comments
17.1	Have employees been adequately trained on the operating and maintenance procedures for the system?	Y	N	N/A	
17.2	Are operating procedures adequately documented?	Y	N	N/A	
17.3	Is there a user/operator/maintenance manual for the system?	Y	N	N/A	
17.4	Does the manual describe:				
	a) Limits of safe operation, and the implications of exceeding them?	Y	N	N/A	
	b) How the system takes the process to a safe state?	Y	N	N/A	
	c) The risk associated with system failures and the actions required for different failures?	Y	N	N/A	
17.5	Are there means to limit access only to authorized personnel?	Y	N	N/A	
17.6	Can all operational settings be readily inspected to ensure they are correct at all times?	Y	N	N/A	
17.7	Are there means to limit the range of input trip settings?	Y	N	N/A	
17.8	Have adequate means been established for bypassing safety functions?	Y	N	N/A	
17.9	When functions are bypassed, are they clearly indicated?	Y	N	N/A	
17.10	Have documented procedures been established to control the application and removal of bypasses?	Y	N	N/A	
17.11	Have documented procedures been established to ensure the safety of the plant during SIS maintenance?	Y	N	N/A	
17.12	Are maintenance procedures sufficient in detail so as not to leave important interpretations or decisions to maintenance personnel?	Y	N	N/A	
17.13	Are maintenance activities and schedules defined for all portions of the system?	Y	N	N/A	
17.14	Are procedures periodically reviewed?	Y	N	N/A	
17.15	Are procedures in place to prevent unauthorized tampering?	Y	N	N/A	

17.16	Are there means to verify that repair is carried out in a time consistent with that assumed in the safety assessment?	Y	N	N/A	
17.17	Are maintenance and operational procedures in place to minimize the introduction of potential common cause problems?	Y	N	N/A	
17.18	Is the documentation consistent with the actual maintenance and operating procedures?	Y	N	N/A	

Section 18: Testing

Item #	Item	Circle a choice			Comments
18.1	Are documented provisions and procedures in place to allow proof testing of all safety functions, including field devices?	Y	N	N/A	
18.2	Are test procedures sufficient in detail so as not to leave important interpretations or decisions to maintenance personnel?	Y	N	N/A	
18.3	Has the basis for the periodic test interval been documented?	Y	N	N/A	
18.4	Are the following items tested:				
	a) Impulse lines?	Y	N	N/A	
	b) Sensing devices?	Y	N	N/A	
	c) Application logic, computations, and/or sequences?	Y	N	N/A	
	d) Trip points?	Y	N	N/A	
	e) Alarm functions?	Y	N	N/A	
	f) Speed of response?	Y	N	N/A	
	g) Final elements?	Y	N	N/A	
	h) Manual trips?	Y	N	N/A	
	i) Diagnostics?	Y	N	N/A	
18.5	Is there a fault reporting system?	Y	N	N/A	
18.6	Are procedures in place to compare actual performance against the predicted or required performance?	Y	N	N/A	
18.7	Are there documented procedures for correcting any deficiencies found?	Y	N	N/A	
18.8	Is calibration of test equipment verified?	Y	N	N/A	
18.9	Are test records maintained?	Y	N	N/A	
18.10	Do test records show:				
	a) Date of inspection/test?	Y	N	N/A	
	b) Name of person conducting test?	Y	N	N/A	
	c) Identification of device being tested?	Y	N	N/A	
	d) Results of test?	Y	N	N/A	
18.11	Are testing procedures in place to minimize the introduction of potential common cause problems?	Y	N	N/A	
18.12	Is failure rate data periodically reviewed and compared with data used during system design/analysis?	Y	N	N/A	

Section 19: Management of Change

Item #	Item	Circle a choice			Comments
19.1	Are there approval procedures which consider the safety implications of all modifications, such as:				
	a) Technical basis for the change?	Y	N	N/A	
	b) Impact on safety and health?	Y	N	N/A	
	c) Impact on operating/maintenance procedures?	Y	N	N/A	
	d) Time required?	Y	N	N/A	
	e) Effect on response time?	Y	N	N/A	
19.2	Are there procedures that define the level of review/approval required depending upon the nature of the change?	Y	N	N/A	
19.3	Has the proposed change initiated a return to the appropriate phase of the lifecycle?	Y	N	N/A	
19.4	Has the project documentation (e.g., operating, test, maintenance procedures, etc.) been altered to reflect the change?	Y	N	N/A	
19.5	Has the complete system been tested after changes have been introduced, and the results documented?	Y	N	N/A	
19.6	Are there documented procedures to verify that changes have been satisfactorily completed?	Y	N	N/A	
19.7	Have all affected departments been appraised of the change?	Y	N	N/A	
19.8	Is access to the hardware and software limited to authorized and competent personnel?	Y	N	N/A	
19.9	Is access to the project documentation limited to authorized personnel?	Y	N	N/A	
19.10	Are project documents subject to appropriate revision control?	Y	N	N/A	
19.11	Have the consequences of incorporating new versions of software been considered?	Y	N	N/A	

Section 20: Decommissioning

Item #	Item	Circle a choice			Comments
20.1	Have management of change procedures been followed for decommissioning activities?	Y	N	N/A	
20.2	Has the impact on adjacent operating units and facilities been evaluated?	Y	N	N/A	
20.3	Are there procedures to maintain the safety of the process during decommissioning?	Y	N	N/A	
20.4	Are there procedures that define the level of authorization required for decommissioning?	Y	N	N/A	

References

1. *Programmable Electronic Systems in Safety Related Applications - Part 2 - General Technical Guidelines.* U.K. Health & Safety Executive, 1987.
2. ISA-84.01-1996. *Application of Safety Instrumented Systems for the Process Industries.*
3. *Guidelines for Safe Automation of Chemical Processes.* American Institute of Chemical Engineers - Center for Chemical Process Safety, 1993.
4. *Out of Control: Why control systems go wrong and how to prevent failure.* U.K. Health & Safety Executive, 1995.

16

CASE STUDY

Chapter Highlights

16.1 Introduction

The case study presented in this chapter is intended to show how the material and techniques outlined in Chapters 1 to 15 can be used to specify, design, install, commission, and maintain a typical safety instrumented system (SIS). The intention is to cover the lifecycle steps presented in Chapter 2. By reviewing the solutions relating to specific issues in this case study, the reader will be able to further clarify the techniques, better understand and appreciate the need for good documentation, and resolve possible misunderstandings. This chapter should serve as an additional guide to those involved in specifying and designing safety systems.

The emphasis of this chapter is on the approach or method, *not* on the problem or the final solution. The case study and the documentation have been simplified so that the focus remains on "how to" rather than "how many."

A Word of Caution

The controls and protective instrumentation discussed in this chapter are oversimplifications offered for study purposes only. They do not in any way reflect what is or what may be required for actual installations. Also, the solutions and the system proposed do not necessarily comply with some internationally recognized standards (e.g., National Fire Protection Association, American Petroleum Institute).

16.2 The Safety Lifecycle and Its Importance

As highlighted in Chapter 2, the ANSI/ISA 84.00.01-2004, Parts 1-3 (IEC 61511 Mod) standard lists the various steps that need to be followed to ensure that safety instrumented systems are specified, designed, installed, and properly maintained. Figure 16.1 is a simplified version of the lifecycle. The various steps can be divided into three main phases:

1. risk analysis
2. design and implementation
3. operation and maintenance

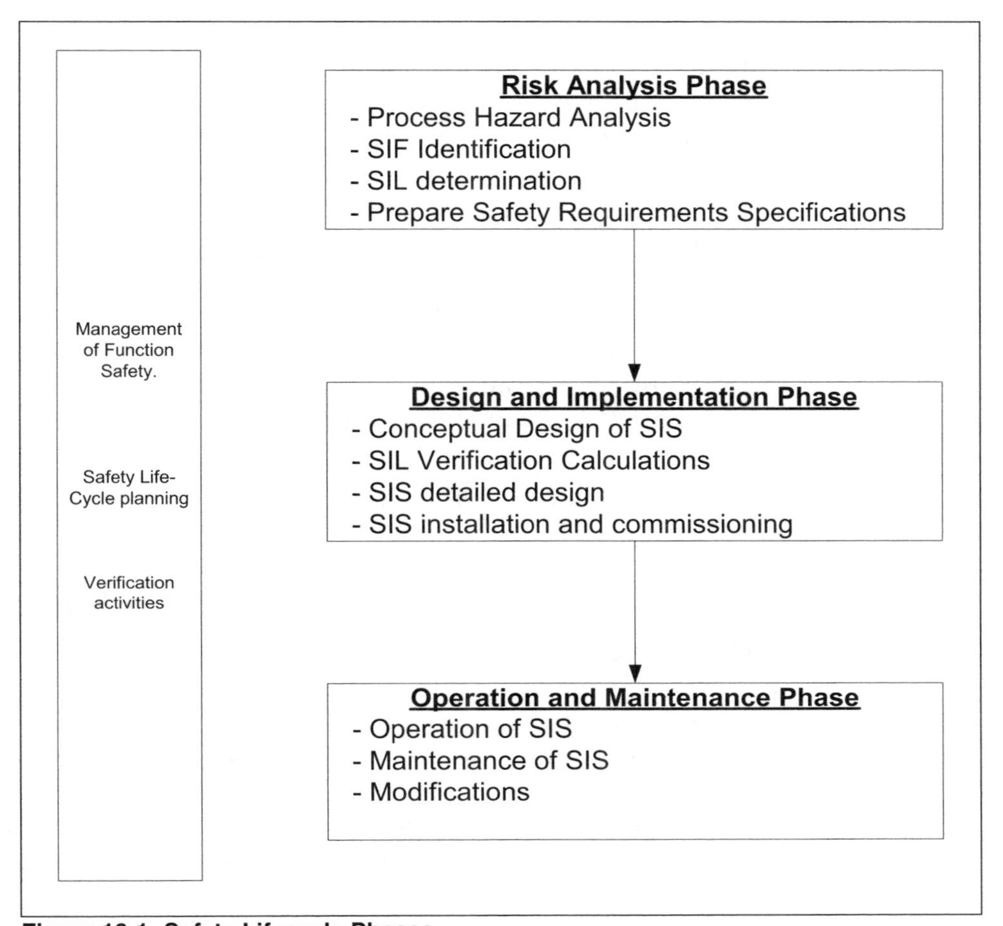

Figure 16-1: Safety Lifecycle Phases

The risk analysis phase covers the process hazards analysis, risk evaluation, identification of safety instrumented functions (SIFs), safety integrity level (SIL) determination, and the safety requirements specification (SRS).

The design and implementation phase covers the design and implementation of the SIS. This includes the SIL verification calculations to verify that the system being installed satisfies the SIL requirements, along with installation and commissioning.

The operation and maintenance phase deals with operating and maintaining the system throughout its life in order to ensure that the SIL for each function is sustained.

The standard also emphasizes that the management of functional safety, functional safety assessments, safety lifecycle structure and planning, and verification activities should occur throughout the lifecycle. This is indicated by the vertical box on the left side of Figure 16-1.

All of the above activities need to be addressed for the following reasons:

- Chapter 2 reviewed the findings of the United Kingdom Health and Safety Executive (U.K. HSE) that showed that accidents due to control and safety system failures could be traced back to errors that could have been caught by adherence to such a lifecycle.
- Systematic failures can be reduced by following the lifecycle. Such failures are caused by inadequate specifications, design, software development, testing, training, and procedures.
- Safety integrity levels need to be recognized as a lifecycle issue. In order to sustain the SIL determined for the various safety functions, all of the lifecycle steps need to be completed and verified.

16.3 Case Description: Furnace/Fired Heater Safety Shutdown System

The safety instrumented system for the fuel controls of a furnace/fired heater, which is part of a crude unit in an oil refinery, needs to be upgraded. Plant management is concerned that the existing systems are inadequate, ineffective, and unreliable. The furnace is natural draft, natural gas fired, with continuous pilots. Figure 16-2 shows a diagram of the furnace and the associated basic process controls.

Description of Basic Process Controls

The feed flow to the furnace is controlled by regulatory control loops FC-9 and FC-10. A local self regulating pressure controller (PCV-6) controls the pilot gas pressure. The fuel gas flow and coil outlet temperature controls are part of a cascade control system. The gas flow controller (FC-3) is in cascade with temperature controller (TC-8). The flow controller manipulates valve FV-3 to control the gas flow at a set point manipulated by the temperature controller. All regulatory control loops are in the basic process control system (BPCS), which is located in a control room approximately 200 feet away from the furnace.

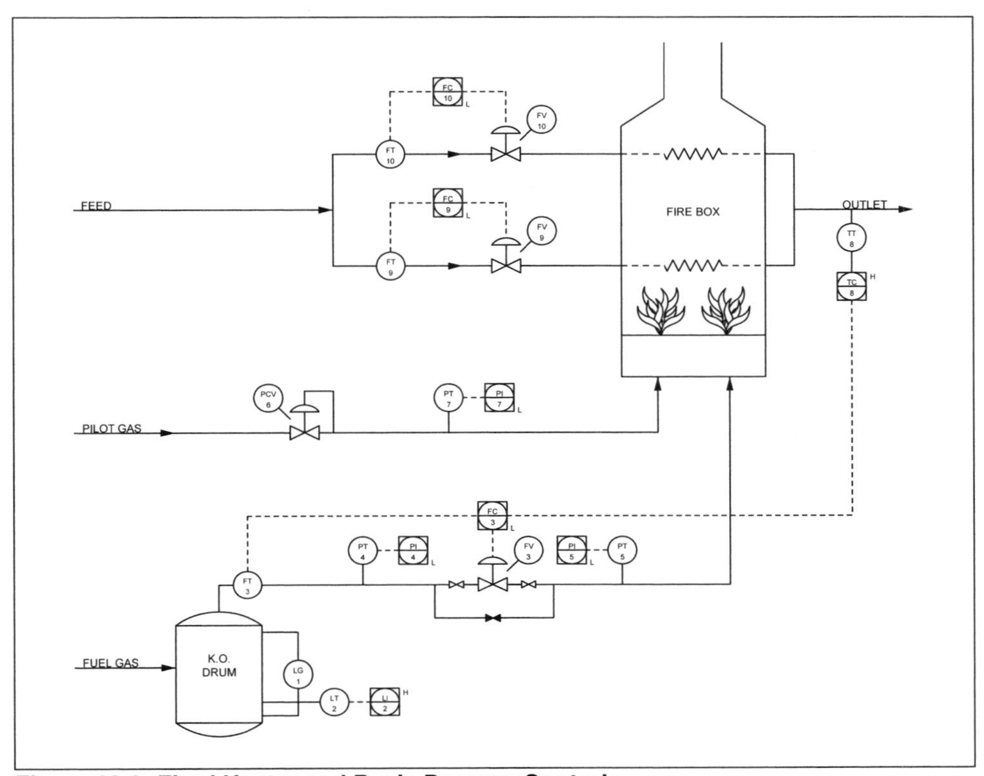

Figure 16-2: Fired Heater and Basic Process Controls

The following hazardous events were identified during the process hazard review:

Table 16-1: Furnace Hazards

Hazard	Possible Cause	Consequence	Likelihood
Furnace Explosion	Loss of pilots	Multiple Fatalities	Medium
Furnace fire	Loss of pilots, or loss of flow in furnace tube	$1,000,000 loss	Medium
Tube failure	Loss of flow in furnace tube	$1,000,000 loss	Medium

It was felt that no additional non-instrumented layers could be applied to prevent or mitigate the above events. The following safety and protective instrumentation was recommended.

1. Fuel gas to be shutoff on low flow on any pass (with 20-sec delay)
2. The fuel gas and pilots to be shutoff on low pilot gas pressure

3. A dedicated hardwired switch to be provided in the main control room for manual shutoff of fuel and pilots

A demand on each safety instrumented function is expected to occur at a frequency less than once per year. Also, one nuisance trip per year of the furnace due to a failure of the safety system can be tolerated. The cost impact of a nuisance trip is approximately $20,000.

16.4 Scope of Analysis

The scope of the analysis includes the following safety lifecycle steps:

1. define the target SIL for each SIF
2. develop the safety requirements specification (SRS)
3. SIS conceptual design
4. lifecycle cost analysis
5. verify that the conceptual design meets the SIL
6. detailed design
7. installation, commissioning, and pre-startup testing
8. establish operation and maintenance procedures

Not included in the above scope are other key lifecycle activities that also need to be completed prior to, or in conjunction with, the above activities. These are:

- Safety lifecycle planning. A safety lifecycle plan needs to be developed and established for the project. The standards do not require that operating companies follow the safety lifecycle as specifically outlined in the standards.
- Functional safety management activities. Some of these include:
    - creating a functional safety management plan
    - defining the roles and responsibilities of personnel
    - addressing personnel competency
    - documentation and documentation control
    - planning for functional safety verification
    - planning for functional safety assessment
    - safety audits
- Verification activities associated with each step of the lifecycle.

- Functional safety assessment. An assessment must at least be carried out prior to the startup of the plant.

16.5 Define Target SILs

A safety integrity level (SIL) must be determined for each safety instrumented function, several methods for SIL determination were outlined in Chapter 6, including:

Qualitative:Using the 3-D risk matrix

QualitativeUsing the risk graph

QuantitativeLayer of protection analysis (LOPA)

The 3-D risk matrix will be used for this case study. The consequence and likelihood of the hazardous events are documented in Table 16-1.

> **A Word of Caution**
>
> **The consequence categories used in the SIL determination are the worst credible consequence assuming that no safeguards exist. The likelihood categories assume that all safeguards exist except the safety instrumented function (SIF) being reviewed. In many instances, the likelihood categories in the safety study and report *include* the SIF. If this is the case, the SIL team should redo the likelihood category assuming that the SIF does *not* exist.**

The risk/SIL determination matrix shown below is a simplification of the matrix shown in Figure 6-2. This matrix shows the SIL required for various consequence and likelihood categories assuming no additional safety layers (as was determined during the safety study).

Consequence				
	Hi	2	3	3
	Med	2	2	3
	Low	1	2	2
		Low	**Med**	**Hi**
		Likelihood		

The SIL required for the two SIFs based on the above matrix is shown in Table 16-2.

Table 16-2: Safety Integrity Level Requirements for Fired Heater Example

Item	Safety Function	Consequence		Likelihood	SIL
1	Low or loss of flow in passes	Tube failure, $1,000,000 loss	Med	Med	2
		Furnace fire, $1,000,000 loss	Med	Med	
2	Low pilot gas pressure	Explosion, multiple fatalities	Hi	Med	3

Note that a SIL was not selected for the manual shutdown. By definition, safety instrumented functions are automated systems that place the equipment or process in a safe state when certain conditions are violated. Alarms and manual shutdowns are usually *not* considered safety instrumented functions since they do not take automatic action. Therefore, many companies do not require SIL selections for alarm and manual shutdowns.

16.6 Develop Safety Requirement Specification (SRS)

Chapter 5 covered the safety requirement specification (SRS), which consists of the functional specification (what the system does), and the integrity specification (how well it does it). The SRS may in reality be a series of documents. Table 16-3 summarizes the key information that should be included in the SRS. Listing all of the requirements in a single table enables a simple crosscheck of what needs to be provided. The "details" column identifies whether information pertaining to each item should be provided for the application. If required, any special comments can be included in this column.

Table 16-3: SRS Summary Table

Item	Details of Requirement
Input Documentation and General Requirements	
P&ID's	Required. Figure 16-2 is a simplification of the actual P&ID.
Cause and effect diagram	Attached. (Refer to Table 16-4)
Logic diagrams	The cause and effect drawing is adequate to identify the logic requirements.

Process data sheets	Need to be provided for all field devices (i.e., FT-22, FT-23, PT-7, PT-24 and the four shutdown valves).
Process information (incident cause dynamics, final elements, etc.) of each potential hazardous event that requires a SIF.	Detailed descriptions are required on: - Details as to how explosions, fires, or tube failures can occur and how the protective instruments will mitigate these occurrences. - Speed of response and accuracy of the protective system needs to be defined. - Special requirements for the shutdown valves (e.g., fire safety, shutoff class).
Process common cause failure considerations such as corrosion, plugging, coating, etc.	The crude flow measurement is a difficult application due to high viscosity and freeze up. An in-line device is preferable. For this application the site has had good experience with Coriolis type meters. Corrosion due to hydrogen sulfide in the atmosphere needs to be addressed.
Regulatory requirements impacting the SIS.	The requirements of ANSI/ISA-84.00.01-2004, Parts 1-3 are to be satisfied. API 556 can be used as a reference. Full compliance with the standard is not mandatory.
Other	None
Detailed requirements for each SIF	
ID number of SIF	Refer to Table 16-4.
Required SIL for SIF	Refer to Table 16-4.
Expected demand rate	Once per year. Low demand operation.
Test interval	3 months
The definition of the safe state of the process, for each of the identified events.	The safe state is to shutoff the fuel gas, pilot gas, and the feed to the furnace.
The process inputs to the SIS and their trip points.	Refer to Table 16-4.
The normal operating range of the process variables and their operating limits.	Refer to Table 16-4.
The process outputs from the SIS and their actions.	Refer to Table 16-4.
The functional relationship between process inputs and outputs, including logic, math functions, and any required permissives.	Refer to Table 16-4.
Selection of de-energized to trip or energized to trip.	The complete safety system shall be de-energized to trip.

Consideration for manual shutdown.	One hardwired shutdown switch to be located in the main control room (HS-21). The manual shutdown must be independent of the programmable system.
Requirements related to the procedures for starting up and restarting the SIS.	Required. To be developed.
List of any special application software safety requirements.	None required.
Requirements for overrides or bypasses including how they will be cleared.	Required. To be developed.
Action(s) to be taken on loss of energy source(s) to the SIS.	All trip valves to close.
Response time requirement for the SIS to bring the process to a safe state.	5 seconds is acceptable.
Response action to diagnostics and any revealed fault.	Immediate response. Maintenance to be assigned highest priority. Closer monitoring of critical parameters must be put in place by operations until the problem is rectified.
Human-machine interfaces requirements.	Dedicated hardwired alarms in the main control room are required for any safety system fault or trip condition. Manual shutdown button required.
Reset function(s).	The trip valves shall be provided with manual reset solenoids.
Requirements for diagnostics to achieve the required SIL.	Smart field sensors are to be provided in order to utilize their diagnostic capabilities. Shutdown valves are to be provided with limit switches to verify that they have closed when requested by the logic system.
Requirements for maintenance and testing to achieve the required SIL.	Refer to the manual testing procedures.
Reliability requirements if spurious trips may be hazardous.	The spurious trip rate is to be calculated for compliance with the safety study requirements.
Failure mode of each control valve.	Fail closed.
Failure mode of all sensors/transmitters	Fail low.
Other.	None.

Table 16-4: Cause and Effect Diagram (Refer to Figure 16-3)

Tag #	Description	SIL	Instrument Range	Shut down Value	Units	Close Valve XV-30A	Close Valve XV-30B	Close Valve XV-31A	Close Valve XV-31B	Notes
FT-22	Pass flow A to furnace	2	0-500	100.0	B/H	X	X			1
FT-23	Pass flow B to furnace	2	0-500	100.0	B/H	X	X			1
PT-7/24	Pilot gas pressure	3	0-30	5	PSIG	X	X	X	X	2
	Loss of control power	NA				X	X	X	X	
	Loss of instrument air	NA				X	X	X	X	
HS-21	Manual Shutdown	NA				X	X	X	X	
Note 1: 20 second delay required before closing valves. Note 2: Solenoids with manual reset to be provided for all four shutdown valves.										

16.7 SIS Conceptual Design

The conceptual design must comply with any relevant company standards. A summary of the company guidelines relating to SIL and hardware selection is:

Table 16-5: SIS Design Guidelines Based on SIL

SIL	Sensors	Logic Solver	Final Elements
3	Redundant sensors required, either 1oo2 or 2oo3 depending on spurious trip requirements.	Redundant safety PLC required.	1oo2 voting required.
2	Redundancy may or may not be required. Initial option is not to have redundancy. Select redundancy if warranted by PFD_{avg} calculations.	Safety PLC required.	Redundancy may or may not be required. Initial option is not to have redundancy. Select redundancy if warranted by PFD_{avg} calculations.
1	Single sensor	Non-redundant PLC or relay logic.	Single device.

Based on the criteria in Table 16-5, the proposed system is shown in Figure 16-3.

Conceptual Design Requirements

The conceptual design (see Chapter 10) builds on and supplements the safety requirement specification (SRS) covered in Section 16.6 and incorporates the company's design guidelines. Key information required by the engineering contractor in order to complete the detailed engineering package should be provided. The design should adhere to the company standards and procedures. There should be no contradiction between the SRS and the conceptual design requirements.

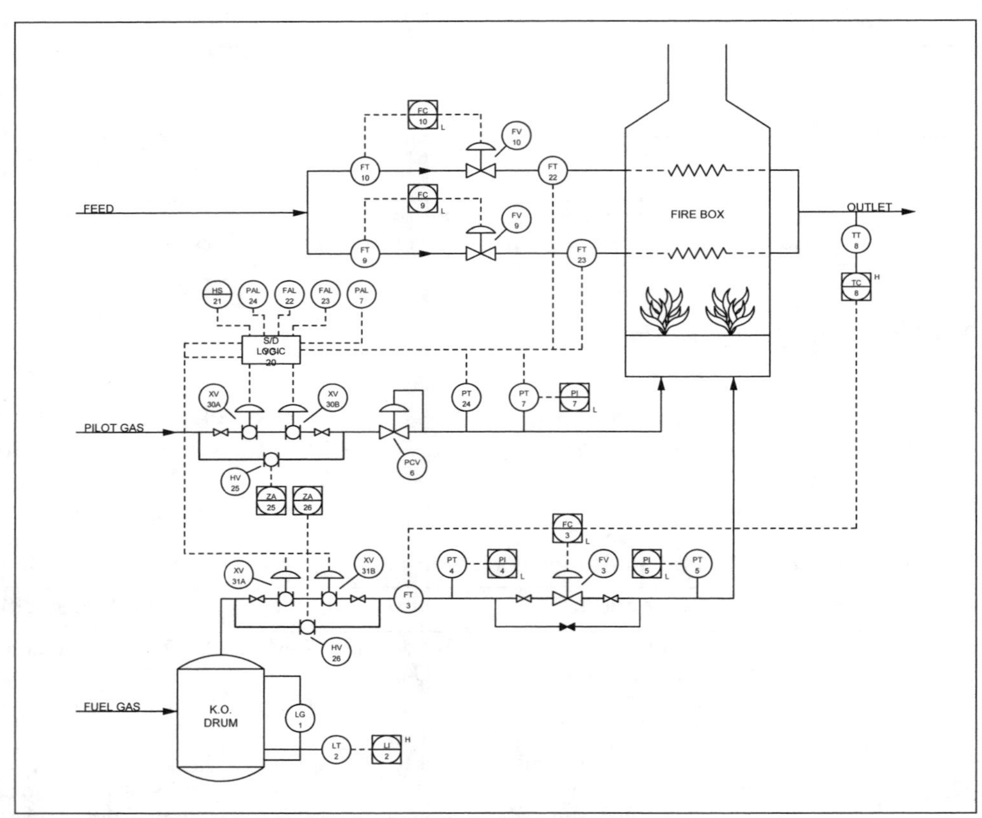

Figure 16-3: Proposed Safety Instrumented Functions

Table 16-6 summarizes the basic conceptual design requirements.

Table 16-6: Conceptual Design Summary

System architecture:	The logic solver shall be a redundant, certified safety PLC. The cabinet shall be located in the main control building. 1oo2 voting is required for the pilot gas pressure transmitters, as well as the fuel gas and pilot gas shutdown valves. Verification is required that this design will meet the SIL requirements. Figure 16-4 is a sketch of the overall configuration.
Minimize common cause:	Wiring from the 1oo2 valves and transmitters to the SIS to be segregated from BPCS wiring. Separate uninterruptible power supply (UPS) to power the SIS. All transmitters to have separate taps.
Environmental considerations:	The area classification is Class 1, Group D, Div 2. Hydrogen sulfide gas is in the environment around the furnace. The ambient temperature can fall to -35°C during the winter.
Power supplies:	110v, 60hz power is available from two separate UPS systems located in the main control room.
Grounding:	Ensure that company grounding standards for instrument and power systems are followed.
Bypasses:	Bypass valves are required to be installed around the pair of trip valves for the pilot gas and fuel gas trip valves for on-line testing. An alarm in the BPCS is required to indicate that a bypass valve has opened. No other bypasses are required.
Application software:	Ladder logic to be used for all programming in the SIS.
Security:	The existing company security requirements for access to and modification of the SIS logic shall be followed.
Operators interface:	Shutdown and diagnostic alarms are to be wired to an existing hardwired annunciator. The bypass alarms are to be connected to the BPCS.

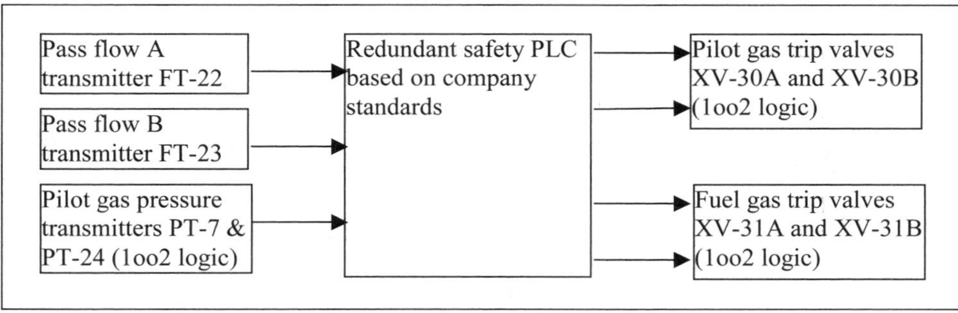

Figure 16-4: Proposed Conceptual SIS Design

16.8 Lifecycle Cost Analysis

Table 16-7: Lifecycle Cost Summary

Lifecycle Costs (20 years)	Material $	Labor $	Total cost $	Subtotal $
Initial Fixed Costs				
Safety classification		$1,000	$1,000	
SRS/Design specifications		$3,000	$3,000	
Detailed design and engineering		$20,000	$20,000	
Sensors	$24,000		$24,000	
Final elements	$6,000		$6,000	
Logic system	$30,000		$30,000	
Misc. – Power, wiring, jb's	$4,000		$4,000	
Initial training		$5,000	$5,000	
FAT/Installation/PSAT	$4,000	$16,000	$20,000	
Startup and correction	$1,000	$2,000	$3,000	
Fixed costs subtotal				$116,000
Annual Costs				
Ongoing training		$1,000	$1,000	
Engineering changes	$1,000	$1,000	$2,000	
Service agreement		$1,000	$1,000	
Fixed operation and maintenance costs		$1,000	$1,000	
Spares	$4,000		$4,000	
Online testing		$8000	$8,000	
Repair costs	$1,000	$500	$1,500	
Hazard costs				
Spurious trips costs			$8,000	
Annual costs subtotal				$26,500
Present value for annual costs (20 years, 5% interest rate)				$330,249
Total Lifecycle Costs				**$446,249**

16.9 Verify that the Conceptual Design Meets the SIL

It's important to verify that each safety instrumented function (SIF) meets the safety integrity level (SIL) requirements. In this case we have three functions (i.e., low pass flow A, low pass flow B, and low pilot gas pressure). The low pilot gas pressure function requires SIL 3 and the pass flow functions require SIL 2. Only the SIL 3 function will be analyzed here.

The equations outlined in Chapter 8 are used for the calculations. A block diagram of the conceptual pilot gas shutdown function is shown in Figure 16-5.

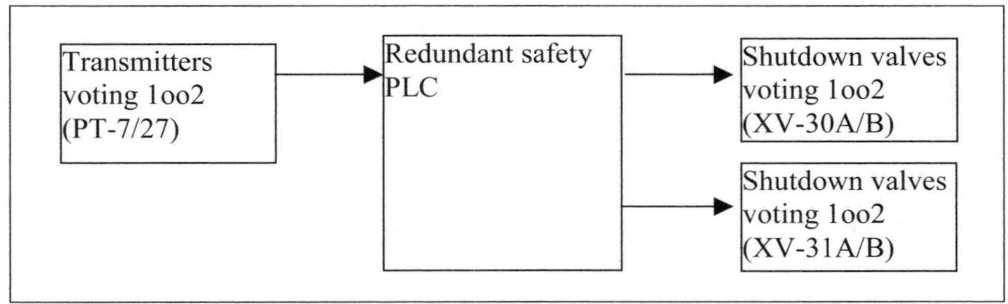

Figure 16-5: Block Diagram for Pilot Gas Shutdown Function

Assuming:

Mean time to repair (MTTR):	8 hours
Average demand rate:	Once per year
Manual test interval :	3 months
Common cause β factor:	5%
Transmitter diagnostics using comparison:	90%

Table 16-8: Failure Rate Data (Failures per Year)

Item	Dangerous failure rate λ_d	Safe failure rate λ_s
Transmitter PT-7/24	0.01 (1/100 years)	0.02 (1/50 years)
Valve and solenoid XV-30A/B, XV-31A/B	0.02 (1/50 years)	0.1 (1/10 years)
Safety PLC	See note 1	See note 1
Note 1: The PFD_{avg} and MTTFsp values for the redundant safety PLC are supplied by the vendor.		

PFD_{avg} Calculations:

PFD_{avg} (Sensors) = 0.01 * 0.1 * 0.05 * 0.25/2 = 0.00000625
(failure rate * 10% undetected failures * 5% common cause * test interval/2)
PFD_{avg} (Valve and solenoid) = 2 * 0.02 * 0.05 * 0.25/2 = 0.00025
(quantity * failure rate * 5% common cause * test interval/2)
Redundant safety PLC: = = 0.00005
PFD_{avg} (Total) = 0.000306

The maximum value allowed for SIL 3 is 0.001, therefore the conceptual design satisfies the safety requirements. The risk reduction factor (RRF = 1/PFD) for the system is 3,300, which is between the range of 1,000 and 10,000 for SIL 3.

$MTTF^{spurious}$ Calculations:

All field devices are included in nuisance trip calculations.

$MTTF^{spurious}$ (Sensors) = 1/(4 * 0.02) = 12.5 yrs
$MTTF^{spurious}$ (Valve and Solenoid) = 1/(4 * 0.1) = 2.5 yrs.
MTBFsp (Safety PLC) = 1/(0.01) = 100 yrs.
MTBFsp (Total) = 2.0 years

A nuisance trip is expected to occur, on average, every two years, which also meets the original requirements.

16.10 Detailed Design

The detailed design needs to comply with the safety requirement specification and the conceptual design. The following documentation shall be provided as part of the detailed design package.

1. results and recommendations of safety/hazard studies
2. all PFD_{avg} and $MTTF^{spurious}$ calculations
3. piping and instrument diagrams (P&IDs)
4. instrument index
5. specification sheets for safety PLC, transmitters, and valves
6. loop diagrams
7. cause and effect matrix
8. drawings to indicate locations of all major equipment and controls

9. PLC system configuration, I/O listings, and ladder program listings
10. junction box and cabinet connection diagrams
11. power panel schedules
12. pneumatic system tubing diagrams
13. spare parts list
14. vendor standard documentation including specifications, installation requirements, and operating/maintenance manuals
15. checkout forms and procedures
16. manual test procedures

16.11 Installation, Commissioning and Pre-startup Tests

The following activities need to be completed.

- Factory acceptance test (FAT) of logic system. The responsibilities and details of the test need to be included as part of the detail design activities. Support personnel shall witness the test.
- Field installation of the complete system. The field installation must conform to the installation drawings included in the design package.
- Installation checks. Device checkout forms must be completed and included with the installation scope of work.
- Validation or site acceptance tests. Functional checkout per the functional checkout forms and procedures.
- Pre-startup acceptance test (PSAT). To be completed by operations personnel prior to startup of the unit. The cause and effect diagram will be adequate for the PSAT.

16.12 Operation and Maintenance Procedures

Procedures must be developed for safe operation and maintenance of the unit. Test procedures must be developed for testing each safety instrumented function. The following procedure covers testing of the low pilot gas pressure shutdown function.

Shutdown Test Procedure — Procedure # XX-1

1.0 Purpose
To test the low pilot gas pressure trips for furnace F-001 and transmitters PT-7 and PT-24.

2.0 Responsibility for test
The furnace operator is responsible for completing the test. The resident instrument technician shall accompany the operator during the test.

3.0 Tools and other items required
Portable radio. No special tools are required to complete this test.

4.0 Trip check frequency
This test has to be completed every 3 months

5.0 Hazards
Failure to open the bypass valves for the pilot gas or the fuel gas valves will trip the furnace and can create a furnace fire or explosion.

Failure to close the bypass valves fully upon completing the test can disable the furnace shutdown functions.

6.0 Reference information
The following documentation should be readily available to assist in rectifying any problems: loop drawings, instrument spec sheets, and shutdown system description.

7.0 Detailed step-by-step test procedure
Refer to attached sketch (Figure 16-6) with trip system description.

Step #	Action	Check OK
1	Advise all operating unit personnel that the test is about to commence.	
2	Open pilot gas bypass valve HV-25 fully. Verify bypass alarm ZA-25 in BPCS. Acknowledge the alarm.	
3	Open fuel gas bypass valve HV-26 fully. Verify bypass alarm ZA-26 in BPCS. Acknowledge the alarm.	
4	Get confirmation that the furnace operation is still stable.	
5	Close process isolation valve for PT-7. Slowly vent the pressure to the transmitter. The two pilot gas trip valves and two fuel gas trip valves should trip at 5 psig. Record trip pressure. Verify that all four valves have closed fully. Verify the speed of operation is < 3 seconds.	

6	Verify that alarm PAL-7 was activated in the main control room annunciator and in the BPCS. Acknowledge the alarm.	
7	Re-pressurize PT-7.	
8	Alarm PAL-7 should clear. Verify	
9	Reset the solenoids for the four trip valves. The valves should reopen instantly. Verify.	
10	Repeat steps 5 through 9 for PT-24.	
11	Close pilot gas bypass valve HV-25 fully. Verify that alarm ZA-25 in the BPCS cleared.	
12	Close fuel gas bypass valve HV-26 fully. Verify that alarm ZA-25 in the BPCS cleared.	
13	Advise all operating personnel that the trip test for PT-7 and PT-24 is complete.	
14	Complete, sign, and forward the completed documentation to plant engineer for review and filing.	

Results/Comments/Deficiencies:

Shutdown Check Completed By:________________ Date: _______

Operation: If either PT-7 or PT-24 pressure drops below 5 PSIG valves XV-30A, XV-30B, XV-31A and XV-31B will close within 3 seconds.

Summary

The fired heater case study in this chapter shows how the material and techniques outlined in Chapters 1 through 15 can be used to specify, design, install, commission, and maintain atypical safety instrumented system. The following sections of the lifecycle model were addressed for the application:

- define the target SIL for each SIF
- develop the safety requirement specification (SRS)
- SIS conceptual design
- lifecycle cost analysis
- verify that the conceptual design meets the SIL
- detail design
- installation, commissioning and pre-startup tests
- operation and maintenance procedures

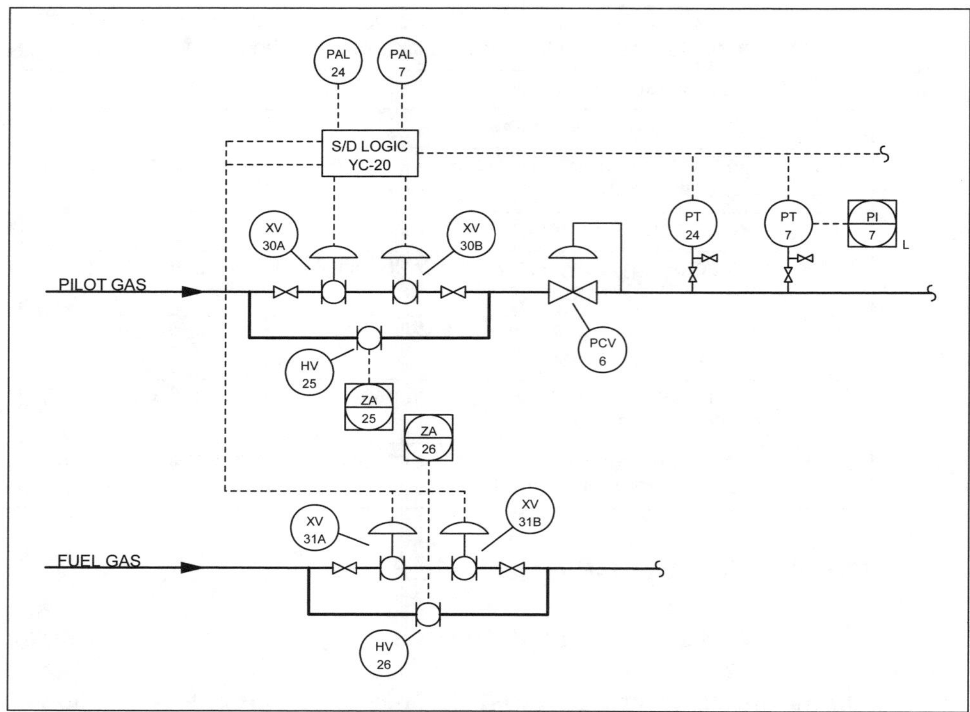

Figure 16-6: Low Pilot Gas Pressure Trip System Diagram and Description

References

1. ANSI/ISA-84.00.01-2004, Parts 1-3 (IEC 61511-1 to 3 Mod). *Functional Safety: Safety Instrumented Systems for the Process Industry Sector.*

INDEX